# New Physics in $b$ Decays

## Recommended Titles in Related Topics

*B Decays*
edited by Sheldon Stone
ISBN: 978-981-02-0708-3
ISBN: 978-981-02-1330-5 (pbk)

*B Decays*
*Revised Second Edition*
edited by Sheldon Stone
ISBN: 978-981-02-1836-2
ISBN: 978-981-02-1897-3 (pbk)

# New Physics in $\bar{b}$ Decays

## Marina Artuso
*Syracuse University, USA*

## Gino Isidori
*University of Zurich, Switzerland*

## Sheldon Stone
*Syracuse University, USA*

World Scientific

NEW JERSEY · LONDON · SINGAPORE · BEIJING · SHANGHAI · HONG KONG · TAIPEI · CHENNAI · TOKYO

*Published by*

World Scientific Publishing Co. Pte. Ltd.

5 Toh Tuck Link, Singapore 596224

*USA office:* 27 Warren Street, Suite 401-402, Hackensack, NJ 07601

*UK office:* 57 Shelton Street, Covent Garden, London WC2H 9HE

**Library of Congress Cataloging-in-Publication Data**

Names: Stone, Sheldon, author. | Artuso, Marina, author. | Isidori, G. (Gino), author.

Title: New physics in *b* decays / Sheldon Stone, Marina Artuso, Gino Isidori.

Description: New Jersey : World Scientific, [2022] | Includes bibliographical references.

Identifiers: LCCN 2021050571 (print) | LCCN 2021050572 (ebook) |
   ISBN 9789811251290 (hardcover) | ISBN 9789811251306 (ebook for institutions) |
   ISBN 9789811251313 (ebook for individuals)

Subjects: LCSH: Beta decay. | CP violation (Nuclear physics) | Standard model (Nuclear physics)

Classification: LCC QC793.5.B425 S76 2022 (print) | LCC QC793.5.B425 (ebook) |
   DDC 539.7/523--dc23/eng/20211201

LC record available at https://lccn.loc.gov/2021050571

LC ebook record available at https://lccn.loc.gov/2021050572

**British Library Cataloguing-in-Publication Data**

A catalogue record for this book is available from the British Library.

For any available supplementary material, please visit
https://www.worldscientific.com/worldscibooks/10.1142/12696#t=suppl

Typeset by Stallion Press
Email: enquiries@stallionpress.com

# Acknowledgments

Marina Artuso and Sheldon Stone thank the U. S. National Science Foundation and Syracuse University for support. Gino Isidori acknowledges the support of the Swiss National Science Foundation, the European Research Council, and the University of Zurich. We all thank many flavor physics oriented theorists and experimentalists for interesting and useful conversations, and numerous collaborations. Marina Artuso thanks Robert Bernstein and Alexey Petrov for the useful conversations in the shared effort in the current U.S. particle physics community planning study. Gino Isidori is particularly grateful to Riccardo Barbieri, Javier Fuentes-Martin, Admir Greljo, Davide Marzocca, and Nicola Serra, whose critical comments and constructive discussions have contributed to shape many of the ideas presented in chapter 4. Special thanks go to Marzia Bordone, Claudia Cornella, and Phillip Urquijo for reading and commenting on the manuscript. Finally we would like to thank Marcello Rotondo for communications on HFLAV averages, Marcella Bona for help with UTfit averages, and Jerome Charles and Stephane Monteil for help on CKMfitter and CKMlive.

# Preface

At the time of publication one of the authors of this book, Sheldon Stone, will not be here to see the last fruit of his labor published. His coauthors would like to dedicate this book to his memory and to share with the readers the passion and vibrancy that he brought to this project.

Throughout his career, Sheldon was always relentlessly seeking manifestations of new physics in $b$ decays. This was pursued through the development of new detector technologies in calorimetry, tracking or hadron identification, and via different experimental approaches. After pioneering $b$ physics at electro-positron colliders, with the CLEO experiment at Cornell, he played a major role in the development of $b$ physics experiments at hadron machines, demonstrating the high potential of this apparently hostile environment in unveiling a largely unexplored $b$-hadron phenomenology. In parallel, he was constantly involved in the exploration of new signatures and new analysis techniques.

The book originates from an idea of Sheldon. Exciting results challenging our understanding of fundamental physics have emerged from $b$-physics experiments in the last few years. He felt it was important to discuss them in a thorough but also accessible way, both experimentally and theoretically. These results are critically analyzed with a healthy skepticism, but also the growing awareness that they could represent a turning point in our understanding of the laws of Nature. Their implications for the future of this field are also discussed.

Writing this book has been a unique experience: it was an honor to work with a scientist with the drive, knowledge, and wit, characteristic of Sheldon. It was a privilege to share this journey with him, and we are grateful for the wonderful collaboration that we developed along the way.

Marina Artuso and Gino Isidori

# Contents

# Chapter 1

# Introduction

## 1.1 The Standard Model and its open problems

The nature and the fundamental interactions of the basic constituents of matter are well described by the Standard Model (SM), the Quantum Field Theory (QFT) that coherently describes weak, strong and electromagnetic forces at the microscopic level. Within QFT, which merges the principles of Quantum Mechanics and Special Relativity, all the elementary particles are described as excitations of quantum fields. Within the SM we have three basic categories of particles, or quantum fields: force carriers, matter constituents, and scalar particles. The force carriers, technically known as *gauge bosons*, are the excitations of the fields responsible for a specific type of interaction: they are the *gluons* ($g$), mediating the strong interaction, the *photon* ($\gamma$), responsible for the electromagnetic interaction, and the $Z$ and $W$ *bosons*, mediating weak interactions. Their nature is determined by the principle of *gauge invariance* and by the gauge symmetry of the model, which is found to be $SU(3)_c \times SU(2)_L \times U(1)_Y$.

Each SM field is characterised by three gauge quantum numbers: color, associated with the $SU(3)_c$ group relevant to strong interactions, weak isospin, and hypercharge, associated with the $SU(2)_L \times U(1)_Y$ group relevant to weak and electromagnetic interactions. The principle of local gauge invariance is nothing but the invariance of physical processes under generalised phase transformations (described by these abstract unitary groups), in each point of space time. This principle specifies completely the number and properties of the force carriers, and their interactions with the matter constituents. The identification of the $SU(2)_L \times U(1)_Y$ group behind the unified description of weak and electromagnetic interactions by Glashow [1], Weinberg [2], and Salam [3], and the discovery of the phenomenon of asymptotic freedom for strong interactions by Gross, Wilczek [4] and Politzer [5], can be considered the cornerstones of this construction.

The matter constituents are described by fermions fields, i.e. fields whose excitations are spin-1/2 particles, and are organised in three *families* or *generations*. Each family contains four types of fermion fields, two quarks and two leptons, with different gauge quantum numbers. The list of the different quark and lepton fields and their quantum numbers, are reported in Table 1.1. Ordinary matter consists essentially of particles of the first family: the up and down quarks (the constituents of atomic nuclei), the electrons, and the electron-neutrinos (abundantly produced by the fusion reactions occurring inside stars). As far as we know, quarks and leptons of the second and third family are identical copies of those in the first family except for their different, heavier, masses. These heavier copies are unstable particles that can be produced in high-energy collisions and decay very fast, via weak interactions, into lighter particles. Why we have three almost identical replicas of quarks and leptons and what is the origin of their different masses, are among the big open questions in physics.

Within the Standard Model, quark and lepton masses are the results of a specific short-distance interaction, the so-called Yukawa interaction, that connects fermions fields to the Higgs field, which is the only scalar field of the theory. Genuine mass terms for the SM fermions, as well as all

Table 1.1: The fermion content of the SM. The left side shows the mass eigenstates of the different particles, with the electric charge indicated as a superscript in units of the absolute value of the electron charge. For each charged particle there exists a corresponding anti-particle with opposite charge. The right side shows the independent quantum fields, with well-defined gauge quantum-numbers. Here the index $a$, running from 1 to 3, is the flavor index of the electroweak eigenstates, reflecting the 3 generations of quarks and leptons (see section 1.1.1). The label $c$, in $SU(3)_c$, stands for color, while $Y$ denotes the weak hypercharge, $Y = 2(Q - T_3)$, where $T_3$ is the third component of weak isospin.

| Mass eigenstates | | | | | $SU(3)_c$ | $SU(2)_L$ | $Y$ |
|---|---|---|---|---|---|---|---|
| | 1$^{\text{st}}$ gen. | 2$^{\text{nd}}$ gen. | 3$^{\text{rd}}$ gen. | $Q_L^a = \begin{pmatrix} u_L^a \\ d_L^a \end{pmatrix}$ | **3** | **2** | $+\frac{1}{6}$ |
| Quarks | $u^{+2/3}$ | $c^{+2/3}$ | $t^{+2/3}$ | $u_R^a$ | **3** | **1** | $+\frac{2}{3}$ |
| | $d^{-1/3}$ | $s^{-1/3}$ | $b^{-1/3}$ | $d_R^a$ | **3** | **1** | $-\frac{1}{3}$ |
| Leptons | $e^{-1}$ | $\mu^{-1}$ | $\tau^{-1}$ | $L_L^a = \begin{pmatrix} \nu_L^a \\ \ell_L^a \end{pmatrix}$ | **1** | **2** | $-\frac{1}{2}$ |
| | $\nu_1^0$ | $\nu_2^0$ | $\nu_3^0$ | $e_R^a$ | **1** | **1** | $-1$ |

the gauge bosons, are forbidden by electroweak symmetry. However, within the SM the Higgs field has a non-vanishing vacuum expectation value that breaks spontaneously the $SU(2)_L \times U(1)_Y$ symmetry. This mechanism [6] leads to non-vanishing masses for the $W$ and $Z$ bosons and, thanks to the Yukawa interaction, also to all quarks and charged lepton masses. The discovery of the Higgs boson in 2012 [7, 8], namely the observation of the excitation of the Higgs field, and the direct measurement of some of its couplings to the SM fermions [9], provide a remarkable confirmation of this mechanism.

However, despite the impressive phenomenological success of the SM in a variety of laboratory experiments covering a large range of energies, there are various convincing arguments which motivate us to consider this model only as the low-energy limit of a more complete theory. For instance, the model does not provide a successful description of the phenomena of dark matter and dark energy, which are necessary ingredients for a description of the Universe. A coherent merging of the SM with the classical theory of General Relativity, which describes with success gravitational interactions, has also not been found yet. Last but not least, the Higgs sector of the SM is unstable with respect to quantum corrections and would require extra degrees of freedom to be screened from potentially large corrections induced by new high-energy dynamics (if any).

A rather unsatisfactory aspect of the SM is also the way the model describes quark and lepton masses and, more generally, what we denote as *flavor structure*, namely what distinguishes the different fermions beside their gauge quantum numbers. Within the SM this distinction occurs only because of the Yukawa interaction, which accounts for the non-vanishing quark and charged-lepton masses. Strong, weak, and electromagnetic forces are completely determined by the necessity of invariance of the theory under specific (local) symmetry transformations, and are each controlled by a unique coupling. On the contrary, the Yukawa interaction does not arise by the requirement of additional symmetry in the theory: it is compatible with the other symmetries, but it does not add more. Actually it breaks some of the global symmetries respected by the other interactions, in a rather peculiar way, and it is controlled by a large number of free parameters. Moreover, the SM (at least in its minimalistic version, based on a renormalizable Lagrangian, see below) predicts massless neutrinos, whereas experiments measure tiny, non-zero masses, thereby providing indisputable evidence for non-SM dynamics. Last but not least, the mass matrices of quarks and leptons determined from experiments exhibit a very peculiar behavior that does not appear to be accidental, with eigenvalues spanning several orders of magnitude and a mixing pattern that is very different for the quark and lepton sectors.

The current view on the SM is that this is only an *effective theory*: a theory with a limited range of validity, especially at high energies, whose completion is still unknown. The evolved model should contain new degrees of freedom and should contain new phenomena, that we generically denote as *New Physics* (NP). The search for NP, which represents the forefront research in particle physics, is carried on along different directions: from the direct searches of new particles in high-energy collisions, to the investigation of astrophysical phenomena. A very promising direction is that of indirect searches for new particles, via precision studies of low-energy phenomena. It is our aim here to describe a specific class of indirect searches, namely those performed via decays of $b$ quarks.

There are at least three general reason that make these processes particularly interesting. First, as we shall show, they offer a unique opportunity to scrutinize in depth the flavor structure of the theory, investigating the mechanism behind the origin of quark and lepton masses. Second, in the last few years a series of deviations from precise SM predictions started to emerge in the experimental study of $b$-quark decays (the so-called $b$-physics *anomalies*): this phenomenon could indeed represent the first hint of some form of NP. Third, $b$ quarks are currently under investigations at different experimental facilities, from which we should expect a large amount of data in the near future.

### 1.1.1 *The flavor structure of the SM and the CKM matrix*

In order to better appreciate the special role of $b$ quarks, it is useful to give a closer look to the SM Lagrangian. This allows us to present a more rigorous definition of what we have so far loosely denoted as the flavor structure of the Standard Model.

The SM has the most general renormalizable Lagrangian that is consistent with the $SU(3)_c \times SU(2)_L \times U(1)_Y$ gauge symmetry, Lorentz invariance, the fermion content listed in Table 1.1, and the assumption that the spontaneous symmetry breaking of the electroweak symmetry is induced by a single Higgs field ($H$) transforming as doublet under $SU(2)_L$. It can be conveniently divided into three parts:

$$\mathcal{L}_{\text{SM}} = \mathcal{L}_{\text{gauge}}(A, \psi) + \mathcal{L}_{\text{Higgs}}(H, A) + \mathcal{L}_{\text{Yukawa}}(H, \psi), \qquad (1.1)$$

corresponding to the different sectors of the theory. The arguments among brackets in the different pieces of the Lagrangian indicate which type of fields they depend on, according to the general classifications into gauge fields ($A$), fermion fields ($\psi$), and scalar fields ($H$). As shown explicitly in appendix A, the structure of each piece is rather simple if expressed in terms of the fields with well-defined gauge transformation properties, rather than in terms of the superposition of these fields describing the different mass eigenstates.

The gauge sector exhibits a large accidental *flavor symmetry*.[1] The five independent types of fermion fields listed in Table 1.1 appear in three copies, characterised by the index $a$ running from 1 to 3. Gauge interactions are invariant under unitary transformations in the space defined by this index, that we denote as flavor space. More precisely, $\mathcal{L}_{\text{gauge}}(A, \psi)$ remains invariant under any of the five independent unitary transformations (i.e. generalised rotations) in flavor space that we can perform on any of the five fermion fields with well-defined gauge quantum numbers (the explicit definition of the flavor symmetry group can be found in appendix B).

Within the SM, what distinguishes the different generations, i.e. what breaks the flavor symmetry, is only the Yukawa interaction. The latter has the following form

$$-\mathcal{L}_{\text{Yukawa}} = (Y_D)_{ab}\, \overline{Q}_L^a H d_R^b + (Y_U)_{ab}\, \overline{Q}_L^a H_c u_R^b + (Y_E)_{ab}\, \overline{L}_L^a H e_R^b + \text{h.c.},\tag{1.2}$$

where the $3 \times 3$ complex matrices $Y_{D,U,E}$ are denoted as Yukawa couplings. They can be made diagonal via bi-unitary transformations of the fermion fields they are coupled to. For instance, by a suitable unitary transformation of $Q_L$ and $d_R$ fields in flavor space,

$$Q_L \rightarrow V_L^D Q_L, \qquad d_R \rightarrow V_R^D d_R,\tag{1.3}$$

where $V_{L,R}^D$ are $3 \times 3$ unitary matrices, we can put in diagonal form the down-type Yukawa coupling:

$$Y_D \rightarrow (V_L^D)^\dagger Y_D V_R^D = \text{diag}(y_d, y_s, y_b).\tag{1.4}$$

---

[1] A precise definition of what is meant by *accidental symmetry* is postponed to the end of this chapter.

The Yukawa eigenvalues $(y_f)$ are in one-to-one correspondence with fermion masses via the relation

$$y_f = \frac{\sqrt{2}m_f}{v},$$

$(1.5)$

where $v \approx 246$ GeV denotes the vacuum expectation value of the Higgs field (see appendix A for more details, including normalization and definition of $H$ and $H_c$).

Since gauge interactions are left invariant by the transformations in Eq. (1.3), they do not lead to any observable effect. However, we have at our disposal only one unitary transformation for the $Q_L$ field, which can be used to diagonalise either $Y_D$ or $Y_U$, and not both. In order to diagonalise both quark Yukawa couplings we need to rotate independently the $u_L$ and $d_L$ components of the same $SU(2)_L$-charged field $Q_L$. This leads to a non-trivial physical effect: when moving to the mass-eigenstate basis for both up- and down-type quarks, we end up with a flavor non-diagonal coupling between these fields and the $W$ boson[2]:

$$\mathcal{L}_{q,W^{\pm}} = -\frac{g}{\sqrt{2}} \, \bar{u}_L^i \gamma^{\mu} V_{ij} d_L^j \, W_{\mu}^+ + \text{h.c.}$$

$(1.6)$

Here $V_{ij}$ denote the elements of the matrix $V_{\text{CKM}} = (V_L^U)^{\dagger} V_L^D$, universally known as Cabibbo-Kobayashi-Maskawa (CKM) matrix [10, 11] which is of fundamental importance to many studies involving heavy quark decays.

A more detailed discussion about the properties of the CKM matrix is presented later on in this chapter. Here we stop to note that by construction $V_{\text{CKM}}$ is unitary and is defined in the mass-eigenstate basis for both up- and down-type quarks. This is why its elements are usually indicated by the mass-labels of the different quark fields:

$$V_{\text{CKM}} = \begin{pmatrix} V_{ud} & V_{us} & V_{ub} \\ V_{cd} & V_{cs} & V_{cb} \\ V_{td} & V_{ts} & V_{tb} \end{pmatrix}.$$

$(1.7)$

Experimentally the elements of $V_{\text{CKM}}$ exhibits a strong hierarchical structure, with the diagonal elements close to unity, and the off-diagonal elements

---

[2]Here, and in the rest of the book, we use $\{a, b\}$ to denote the flavor indices of the electroweak eigenstates i.e. the fields with well-defined gauge transformation properties (Table 1.1 right), whereas $\{i, j\}$ are used to denote the different quark mass eigenstates (Table 1.1 left).

very suppressed:

$$|V_{us}|, \ |V_{cd}| \approx 0.22, \quad |V_{cb}|, \ |V_{ts}| = \mathcal{O}(|V_{us}|^2), \quad |V_{ub}|, |V_{td}| = \mathcal{O}(|V_{us}|^3).$$
$$(1.8)$$

The origin of this hierarchical structure, as well as the one observed in the fermion masses,

$$
\begin{array}{lll}
m_u \approx 2\,\mathrm{MeV}, & m_c \approx 1.3\,\mathrm{GeV}, & m_t \approx 173\,\mathrm{GeV}, \\
m_d \approx 4\,\mathrm{MeV}, & m_s \approx 95\,\mathrm{MeV}, & m_b \approx 4.2\,\mathrm{GeV}, \\
m_e \approx 0.511\,\mathrm{MeV}, & m_\mu \approx 105\,\mathrm{MeV}, & m_\tau \approx 1.78\,\mathrm{GeV},
\end{array}
\tag{1.9}
$$

has no explanation within the SM.

Summarizing, the flavor structure of the SM is characterized by a large flavor symmetry, defined by the gauge sector, and a well-defined set of symmetry breaking breaking terms, encoded by the Yukawa couplings. The latter leads to the observed parameters in Eqs. (1.8)–(1.9). The non-trivial pattern in these observed values is suggestive of a deeper ultraviolet structure, behind the SM Lagrangian, needed to explain them.

## 1.2 A first overview of $b$ physics

A few key properties make the $b$ quark a special player in testing the flavor structure of the Standard Model and searching for NP. First, the $b$ quark is the heaviest among down-type quarks and can decay to all the four lighter quarks, probing transitions from the 3$^{\mathrm{rd}}$ generation to both the 2$^{\mathrm{nd}}$ and the 1$^{\mathrm{st}}$ one. The smallness of the CKM elements $|V_{cb}|$ and $|V_{ub}|$, and the smallness of the $b$-quark mass ($m_b$) compared to the $W$ mass ($m_W$), ensures that all these transitions are suppressed within the SM, hence are potentially more sensitive to NP. This is to be contrasted to the top quark, which decays predominantly only to the $b$ quark via an unsuppressed amplitude proportional to $|V_{tb}| \approx 1$, further enhancing the decay width by its large mass. In practice, the combination of the lack of CKM suppression and its large mass (compared to the $W$ mass) conspires to have the $t$ quark lifetime too short to form hadrons, precluding many studies. Second, the $\sim$1.5 ps $b$ quark lifetime permits the formation of quasi-stable bound states with other quarks via strong interactions. We call these "$b$ hadrons." Having to deal with a well-defined long-lived set of hadrons simplifies several experimental and theoretical aspects of $b$ physics. This is again to be contrasted to the the top quark, which decays before hadronizing.

Third, the $b$-quark mass is significantly larger with respect to the scale where strong interactions become non-perturbative ($m_b \gg \Lambda_{\mathrm{QCD}}$): this turns out to be a key ingredient to obtain precise predictions for several observables in $b$-hadron decays, which is a pre-requisite for making precise tests of the SM. The $b$ hadrons we will discuss most in this book are the so-called $B$ and $\bar{B}$ mesons, shown in Table 1.2. They are pseudoscalar states formed by an anti-$b$ quark (or $b$ quark) and one the three light quarks (or the corresponding anti-quark), and are the lightest mesons with the given valence structure. We also list the other ground state mesons and $b$-baryons, the quark content, masses, and lifetimes.

There are several amplitudes of interest in $B$ meson decays. The simplest are charged-current transitions involving a single virtual $W^-$, which occur already at the tree-level according to the SM Lagrangian. The corresponding amplitudes, shown as Feynman diagrams in Fig. 1.1 (for a $\bar{B}$ meson), are called *spectator*, color-suppressed spectator, or *annihilation* diagrams. The terminology reflects the role of the companion quark in the $\bar{B}$ meson. The spectator diagram is similar to that of muon decay: here the $W^-$ is emitted together with a $c$ or $u$ quark, and the light anti-quark ($\bar{q}$), i.e. the spectator quark, is only involved with pairing up with the $c$ or $u$ quarks to form one or more hadrons. The final state can be both semileptonic, i.e. include a $\ell^- \bar{\nu}$ pair, or purely hadronic. The color-suppressed diagram, which is present only for pure hadronic transitions, is suppressed compared to the spectator diagram since the quark-anti-quark pair emitted from the $W^-$ must have the same color (anti-color) of the initial valence quarks in

Table 1.2: The ground state $b$-hadrons [12]. There are also antiparticles for each entry.

| Name | $B^0$ | $B^+$ | $B^0_s$ | $B^+_c$ | $\Lambda^0_b$ | $\Xi^0_b$ | $\Xi^-_b$ | $\Omega^0_b$ |
|---|---|---|---|---|---|---|---|---|
| Valence quarks | $\lvert \bar{b}d \rangle$ | $\lvert \bar{b}u \rangle$ | $\lvert \bar{b}s \rangle$ | $\lvert \bar{b}c \rangle$ | $\lvert bud \rangle$ | $\lvert bsu \rangle$ | $\lvert bsd \rangle$ | $\lvert bss \rangle$ |
| Mass (MeV) | 5280 | 5279 | 5367 | 6275 | 5620 | 5792 | 5797 | 6046 |
| Lifetime ($10^{-12}$ $s$) | 1.52 | 1.64 | 1.62 | 0.51 | 1.46 | 1.48 | 1.57 | 1.64 |

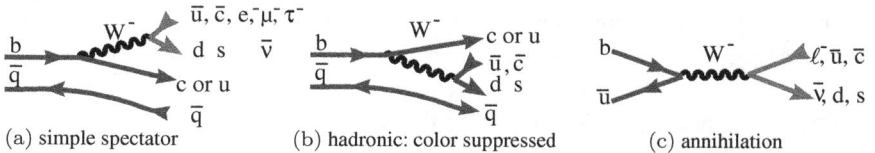

(a) simple spectator     (b) hadronic: color suppressed     (c) annihilation

Fig. 1.1: Spector (a), color-suppressed (b), and annihilation (c) tree-level amplitudes in $\bar{B}$ decays. The allowed disintegrations of the virtual $W$ are shown as the quarks in a $2 \times 1$ column, e.g. $\bar{u}$ over $d$ in (a).

order to form a color-neutral final state. Finally, the annihilation diagram, which can be purely leptonic or purely hadronic, is suppressed both by the small probability of annihilating the two quarks inside the $\overline{B}$ meson and the smallness of $|V_{ub}|$. As we shall discuss in more detail in chapter 2, precise measurements of the semileptonic processes in Fig. 1.1(a) is the way to determine the value of both $|V_{cb}|$ and $|V_{ub}|$ [12].

A very different class of amplitudes are the second-order weak diagrams illustrated in Fig. 1.2, denoted as *box* diagrams, or $|\Delta B| = 2$ amplitudes, which are responsible for the phenomenon of meson-antimeson mixing. This mixing gives rise to damped oscillations in the time-dependent decay distributions of neutral $B(\overline{B})$ mesons to specific final states, which can be measured very precisely obtaining information on both the real and the imaginary parts of the mixing amplitudes. We will discuss this in great detail chapter 2, for both $\overline{B}^0$ and $\overline{B}^0_s$ meson systems. This phenomenon provides both a way to determine the CKM elements $|V_{td}|$ and $|V_{ts}|$, as well as the CP-violating phase of the SM, but is also an extremely powerful probe of physics beyond the SM. Actually the measurements performed so far, being consistent with the SM expectations, have ruled out many motivated NP models.

Another very interesting class of second-order weak diagrams are those responsible for $|\Delta B| = 1$ flavor-changing neutral-current (FCNC) processes, such as the quark-level transitions $b \rightarrow s\gamma$. The first observation of FCNC transitions in $B$ decays has been the CLEO collaboration, who provided evidence for the exclusive reaction $B \rightarrow K^*\gamma$ [13], and later measured also the inclusive process $b \rightarrow s\gamma$ [14]. Representative diagrams for these processes, often called *penguin* diagrams,[3] are shown in Fig. 1.3. Similarly to the box amplitudes mentioned above, these rare processes are

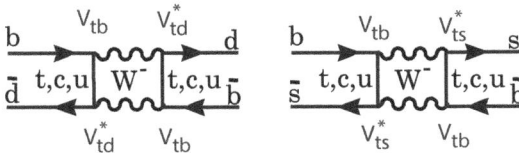

Fig. 1.2: "Box" diagrams illustrating transitions between $\overline{B}^0_{d,s} \rightarrow B^0_{d,s}$ mesons. The relevant CKM elements for the dominant virtual $t$ quarks are shown.

---

[3]While to some these diagrams can appear similar to real penguins, it is worth mentioning that the use of this term resulted from a bet between John Ellis and Melisa Franklin: John lost the bet and had to include the word "penguin" in his next paper.

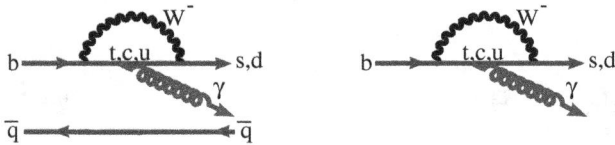

Fig. 1.3: Radiative "Penguin" diagrams relevant to some rare weak decays. The left-side illustrates the exclusive $B \to K^* \gamma$ process where the $s$ quark combines with the $\bar{q}$ spectator quark to form a $K^*$ meson ($\bar{q}$ represents all quarks lighter than the $b$). On the right-side the inclusive process $b \to s\gamma$ is illustrated, where the final hadron state is unmeasured. In both cases the photon can emerge from any charged particle line. The $b \to d$ transition is allowed but suppressed.

extremely powerful probes of NP. Just to mention one concrete example, the comparison of the current measurement of the inclusive $b \to s\gamma$ rate [12] with respect to its theoretical prediction within the SM [15], allow us to set a lower limit on a possible charged Higgs boson mass, within two-Higgs-doublet models, of about 800 GeV. In large areas of the parameter space of this NP model, this limit exceeded the one obtained by direct searches at the LHC. The present frontier on rare FCNC decays is represented by processes mediated by $b \to s\ell^+\ell^-$ amplitudes, which indicates tantalizing hints of deviations from the SM predictions. These will be discussed in great detail chapter 3.1.

### 1.2.1 *Searching for NP in b decays*

So far we highlighted some key properties of the $b$-quark decay amplitudes and their sensitivity to physics beyond the SM. In general, the strategy to search for NP in $b$ decays is based on the following three main objectives:

  **i.** determine precisely the free parameters of the SM Lagrangian;
 **ii.** perform precise data-theory comparisons in SM-allowed processes potentially sensitive to physics beyond the SM;
**iii.** search for SM-forbidden processes.

In this book we will concentrate mainly on the points **i** and **ii** above, and will provide examples for the point **iii**.

    The first objective is a prerequisite for most NP searches. The key point to notice is that the structure of the CKM matrix is largely determined by $b$-quark observables, but it also represents a key input for making precise predictions of virtually any observable in $b$ physics. This is why we devote the first half of chapter 2 to the determination of the CKM elements $|V_{cb}|$ and $|V_{ub}|$, and the measurement of the non-trivial CP-violating phase of the CKM matrix (see section 1.4.1). Strictly speaking this is not a way to search for NP. However, without this knowledge, many NP searches are not feasible. Moreover, comparing precise determinations of the same SM quantity, such as the CKM elements, using independent processes (i.e. via different transition amplitudes) provides an indirect way to search for physics beyond the SM.

As far as the second objective is concerned, we basically concentrate on two sectors: the meson-antimeson mixing amplitudes, discussed in the second half of chapter 2, and decays with dileptons in the final state as well as semileptonic decays, discussed in chapter 3 and 4. The high precision with which we are able to both measure and compute these processes, together with their high-sensitivity to physics beyond the SM, make them extremely interesting probes of non-standard dynamics. As we shall show, with these processes we are able to explore physics occurring around or even above the TeV scale, i.e. well above the $b$ quark mass characterizing the energy of the initial and final states observed in experiments. This is because, even if heavy, NP could leave non-trivial imprint in these rare amplitudes via virtual effects. For instance, the tiny values of the second-order weak processes in Figs. 1.2 and 1.3 make them competitive with first-order NP amplitudes due to a new heavy mediator. The general theoretical tools necessary to describe with high precision these effects, separating the scales of the problem, namely the effective Lagrangians, will be presented later in Section 1.4.

Concerning the last category, in chapter 5 we will focus on lepton flavor violating reactions, in particular $B^0_{(s)} \to \tau^\pm \mu^\mp$ and $B^+ \to K^\mp \tau^\pm \mu^\mp$, as well as processes which violate total lepton number, such as $B^+ \to \pi^- \mu^+ \mu^+$, discussing the measured limits and their implications for physics beyond the SM.

## 1.3 Experimental aspects

Decays of $b$-hadrons are currently examined using $e^+e^-$ colliders operating at center-of-mass energies in the vicinity of the $\Upsilon(4S)$ and at the LHC using $pp$ collisions with center-of-mass energies between 7-13 TeV. Studies at these facilities follow older $\Upsilon(4S)$ $e^+e^-$ experiments such as ARGUS [16] and CLEO [17, 18], higher center-of-mass energy $e^+e^-$ experiments operating at the PEP, PETRA, and LEP accelerators,[4] and the hadron collider experiments CDF [19] and D0 [20].

---

[4]The experiments were PEP: DELCO, HRS, MAC, Mark II, and TPC; PETRA: Mark-J, JADE, and, TASSO; LEP: ALEPH, DELPHI, L3, and OPAL.

We start by first discussing studies at the $\Upsilon(4S)$ resonance from both the BaBar and Belle and experiments. The Belle experiment [21] at the KEK laboratory in Tsukuba, Japan and the BaBar experiment [22] at SLAC in Stanford, Ca. used $e^+e^-$ colliders operating at center-of-mass energies in the vicinity of the $\Upsilon(4S)$ mass using asymmetric beam energies so that the $B$ mesons produced moved along the beam line and thus time dependent measurements were possible. The cross-section for $\Upsilon(4S)$ production is 1.1 nb and the resonance decays into an approximately equal number of $B^-B^+$ and $B^0\overline{B}^0$ mesons [23]. At the peak energy of the $\Upsilon(4S)$, BaBar accumulated an integrated luminosity of 0.433 ab$^{-1}$, while Belle collected 0.771 ab$^{-1}$. Belle using the KEKB accelerator ran at a peak luminosity of $2 \times 10^{34}$ cm$^{-2}$s$^{-1}$. An upgraded experiment, Belle II [24], is matched with the upgraded SuperKEKB accelerator whose goal is to reach peak luminosities of about $6.5 \times 10^{35}$ cm$^{-2}$s$^{-1}$ around the year 2028 (see chapter 6). The detectors are cylindrical with solenoidal magnetic fields. From the inside out they contain silicon vertex detectors, tracking from drift chambers, particle identification,[5] CsI(Tl) crystal electromagnetic calorimeters, based on the CLEO II design [17, 25], a solenoidal magnet, and iron mixed with detection devices to detect $K_L$ mesons and identify muons. The Belle II detector is shown in Fig. 1.4(a); it is quite similar in structure to the Belle detector, and the BaBar detector.

There are several advantages to these experiments: 1) $B$ mesons are approximately 1/4 of the total $e^+e^- \rightarrow$hadrons cross-section, a good signal to background ratio to begin with. When $B$ mesons are produced there are only two of them present; i.e. there are no excess tracks and no extra $e^+e^- \rightarrow$ hadrons interactions, although there can be low energy electromagnetic backgrounds from other beam-beam interactions in the same bunch crossing. 2) The detector acceptance approaches the maximum of $4\pi$ and is mostly quite uniform. 3) The efficiency to learn the flavor of a neutral $B$ meson when studying its partner in the same $\Upsilon(4S)$ decay is quite high, approaching $\sim 40\%$. This is very important for studies of time dependent $CP$ violation. 4) Electrons are easy to find and the small amount of material in the detectors allows them to be measured almost as well as muons. The disadvantages include (1) the relative poor time resolution of $\approx 900$ fs on the decay time difference of two $B$ mesons compared to the $B$ lifetime of $\approx 1500$ fs. This results from a fundamental compromise that is

---

[5]Experiments use various technologies to measure the particles velocity. Since they measure the momentum by bending the tracks in magnetic field, they can determine the particles mass.

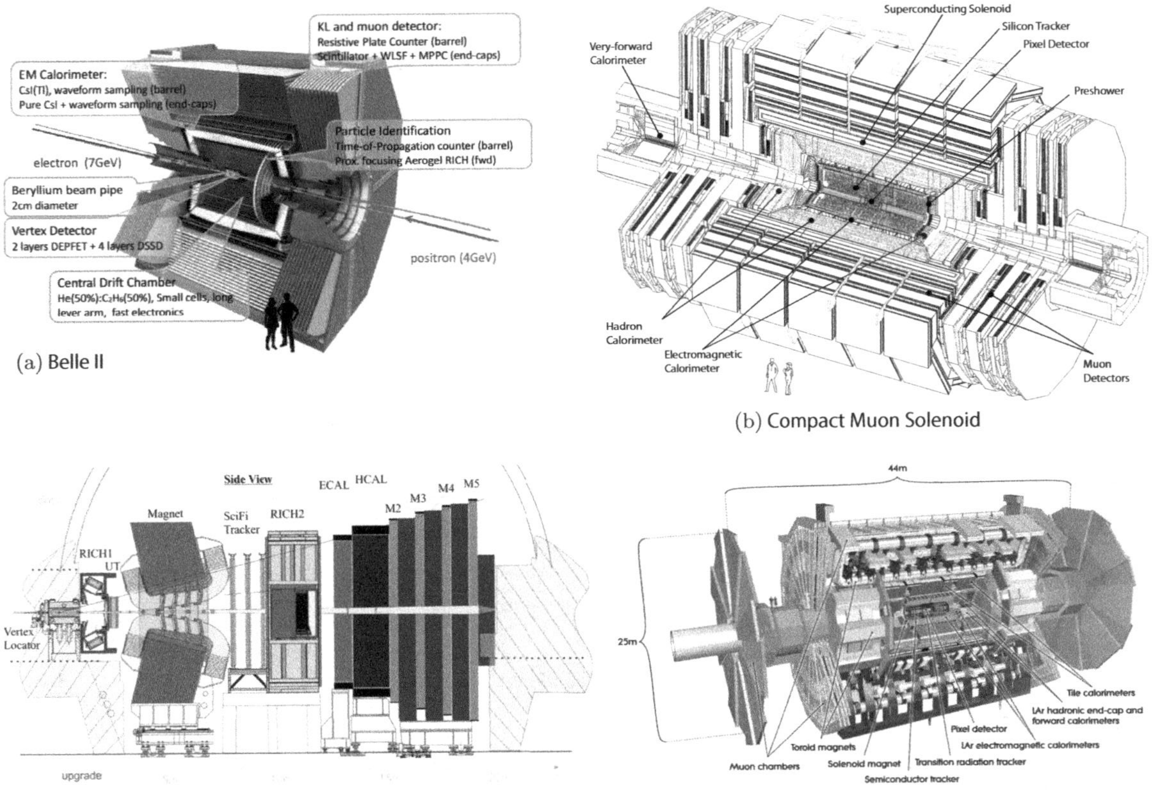

(a) Belle II

(b) Compact Muon Solenoid

(c) LHCb Upgrade I

(d) ATLAS

Fig. 1.4: Sketches of the currently active detectors of experiments contributing to precision studies in *b*-hadron decays: (a) Belle II, (b) CMS, (c) LHCb Upgrade I, and (d) ATLAS. The Belle II and LHCb Upgrade I detectors are similar to their predecessors.

made between having large beam energy asymmetries, which would result in better time resolution and keeping fewer particles from escaping the detector near the beam pipe. The relatively poor decay time resolutions means that it is difficult to tag $B$ mesons by seeing their decays displaced from the primary interaction vertex (PV). (2) Although the luminosities are very high, the cross-section is only 1.1 nb, resulting in relatively small samples especially for "rare" decays. (3) Only $B^- B^+$ and $\overline{B}^0 B^0$ can be studied with any precision. Although $B_s^0$ mesons are produced at the $\Upsilon(5S)$ resonance the cross-section is only about 0.06 nb [26], they are accompanied with other $B$ meson "backgrounds" and the time resolution precludes studies of mixing and $CP$ violation for these decays. (4) Studies of other $b$-hadron species are impossible as the accelerators do not have enough energy to produce them.

Studies of $b$-hadron decays at $pp$ colliders are currently being carried out by the ATLAS [27], CMS [28], and LHCb [29] collaborations using $pp$ collisions at 7, 8 and 13 TeV. The accelerator will likely run at 13.5 TeV when it starts up in 2022. The ATLAS and CMS detectors (see Figs. 1.4(b)(d)) were designed to be sensitive to Higgs boson decays and new high mass particles corresponding to physics beyond the SM. They are cylindrical in geometry, similar in that respect to the $e^+ e^-$ detectors, but lack the ability of distinguishing pions, kaons, and protons. The LHCb detector (see Fig. 1.4(c)), on the other hand is built along the direction of one the beam protons. Because of the kinematics of $b\bar{b}$ production a large number of events with both the $b$ and the $\bar{b}$ particles in the acceptance can be accumulated covering only a small portion of the solid angle. The $b\bar{b}$ production cross-section was measured by the LHCb collaboration to be $72 \pm 0.3 \pm 6.8 \, \mu b$ at 7 TeV and $144 \pm 1 \pm 21 \, \mu b$ at 13 TeV in their detector acceptance. Extrapolating to $4\pi$ using particle production simulations the cross-sections become $\sim 295 \, \mu b$ and $\sim 560 \, \mu b$ for the lower and higher energies [30]. So the LHCb detector accepts $\sim 1/4$ of the $b\bar{b}$ events, while ATLAS and CMS could accept almost the entire cross-section based on geometrical considerations. However, currently none of the LHC detectors can output all the data from all the collisions. Specific selections, called "triggers" are needed for decays of interest. Generally, there are fast hardware triggers followed by further data reduction in software.[6]

---

[6]The upgrade of the LHCb detector currently planned to be installed in late 2021 eliminates the hardware trigger and will allow most $b$-hadron decays of interest to have >90% efficiencies for particles in the acceptance [31].

The largest advantages of the ATLAS and CMS experiments are the large $b\bar{b}$ cross-section coupled with the high luminosity provided by the LHC. Peak luminosities around $2 \times 10^{34} \mathrm{cm}^{-2}\mathrm{s}^{-1}$ are common. The detectors also have excellent identification of muons and electrons. The tracking of the hundreds of particles produced in a single $pp$ collision is very efficient and with the vertex detector clear decays of $b$-flavored hadrons can be seen by their decay products coming from new vertices downstream of the primary collision. ATLAS and CMS separately accumulated $\sim$189 fb$^{-1}$ of data, most of which, $\sim$160 fb$^{-1}$ was accumulated at a center-of-mass energy of 13 TeV. However, they have significant disadvantages for the analysis of $b$-hadron decays. The readout bandwidth of the detectors is severely limited so that $b$-hadron decays with low transverse momentum ($p_T$) or low momentum ($p$) cannot be selected and readout. Since the $p_T$ distribution for $b$-hadrons peaks at half the mass of the $b$-hadron, more $b$'s are lost than acquired. Another obstacle is that the number of interactions in each beam crossing increases with instantaneous luminosity, ranging from about 10 in 2010 to about 40 in 2018.

The LHCb experiment is the only one designed to investigate the decays of beauty and charm hadrons at a hadron collider. Particle production is uniform in the azimuthal angle perpendicular to the beam line, $\phi$, and highly non-uniform in the particles production angle, $\theta$, with respect to the beam line. This is due to a large number of particles concentrating close to the protons directions. A more uniform distribution can be found by using the "pseudorapidty", $\eta$, where $\eta = -\ln\tan(\theta/2)$. Having $b$-hadrons produced somewhat uniformly in $\eta$ means that a detector arrayed along the beam line can capture many of the them by having a much smaller detector than over all of $4\pi$. It also allows for use of two particle identification devices, differing in momentum range sensitivity that form a key ingredient in rejecting backgrounds in the reconstruction of $b$-hadron decays. In the detector acceptance $b$-hadrons are produced with large momenta allowing for significant Lorentz boost. Thus the $b$'s travel on the order of a centimeter before they decay. The resolution on the separation of the $b$ decay vertex from the PV expressed in terms of time is $\approx$40 fs, similar but slightly better than ATLAS and CMS. The only reason that the LHC detectors can detect $b$ hadron decays is because of this precise vertex reconstruction. Often the $b$ decays into a charmed particle. Then the decay chain can have two vertices one from the initial $b$ decay and the other from the subsequent $c$ decay.

A disadvantage of LHCb with respect to ATLAS and CMS is the maximum luminosity the detector can run at. From 2012-2018 LHCb ran at a luminosity of $1.5 \times 10^{33} \text{cm}^{-2}\text{s}^{-1}$, a factor of about 10 less than the other experiments. The number of interactions per crossing was only about 1.5, and the luminosity was generally kept constant during the run by separating the beams. Upgrade I will allow an increase in LHCb luminosity by about a factor of 5.

## 1.4   Theoretical tools

In this section we introduce some basic ingredients necessary to describe, theoretically, processes involving $B$ mesons. We start providing a closer look to the CKM matrix and its different parameterizations. We then proceed introducing the method of effective Lagrangians, which is a very efficient tool to compute of $B$-meson decay amplitudes within and beyond the SM.

### 1.4.1   *Some properties of the CKM matrix*

As shown in section 1.1.1, the CKM matrix is the $3 \times 3$ unitary matrix resulting from the product of the two unitary matrices acting on left-handed quarks necessary to make diagonal both $Y_U$ and $Y_D$. The miss-match of these two diagonalization procedures (in the left-handed sector) is the reason why $V_{\text{CKM}} \neq 1$.

In general, a $N \times N$ unitary matrix depends on $N(N-1)/2$ rotation angles (real parameters) and $N(N+1)/2$ complex phases. However, $2N-1$ phases in $V_{\text{CKM}}$ are not observables since they can be eliminated by relative field redefinitions, i.e. by changing the relative phases of the $u_L^i$ and $d_L^i$ fields. As a result, setting $N = 3$, we deduce that the observable parameters in $V_{\text{CKM}}$ are 3 rotational angles $(\theta_{ij})$ and one complex phase $(\delta)$.

The standard parametrization of Eq. (1.7) adopted by the Particle Data Group [12] is the one proposed by Chau and Keung [32]:

$$V_{\text{CKM}} = \begin{pmatrix} c_{12}c_{13} & s_{12}c_{13} & s_{13}e^{-i\delta} \\ -s_{12}c_{23} - c_{12}s_{23}s_{13}e^{i\delta} & c_{12}c_{23} - s_{12}s_{23}s_{13}e^{i\delta} & s_{23}c_{13} \\ s_{12}s_{23} - c_{12}c_{23}s_{13}e^{i\delta} & -s_{23}c_{12} - s_{12}c_{23}s_{13}e^{i\delta} & c_{23}c_{13} \end{pmatrix},$$

$$(1.10)$$

where $c_{ij} = \cos\theta_{ij}$ and $s_{ij} = \sin\theta_{ij}$ $(i, j = 1, 2, 3)$. Given the strong hierarchical pattern of the off-diagonal elements pointed out in Eq. (1.8), a convenient expanded form of the CKM matrix, often employed in phenomenological analyses, is the parameterization proposed by Wolfenstein in 1983 [33]. Expanding all the elements in powers of the small parameter

$\lambda = |V_{us}| \approx 0.2$, and truncating the expansion up to $\mathcal{O}(\lambda^3)$, leads to [33]

$$V_{\text{CKM}} = \begin{pmatrix} 1 - \lambda^2/2 & \lambda & A\lambda^3(\rho - i\eta) \\ -\lambda & 1 - \lambda^2/2 & A\lambda^2 \\ A\lambda^3(1 - \rho - i\eta) & -A\lambda^2 & 1 - A^2\lambda^4/2 \end{pmatrix} + \mathcal{O}(\lambda^4), \quad (1.11)$$

where $A$, $\varrho$, and $\eta$ are parameters of order 1.

The Wolfenstein parametrization is certainly more transparent than the standard parametrization. However, if one requires sufficient level of accuracy, the terms of $\mathcal{O}(\lambda^4)$ [34] and $\mathcal{O}(\lambda^5)$ have to be included in phenomenological applications. This can be achieved in many different ways, according to the convention adopted. The simplest (and nowadays commonly adopted) choice is obtained *defining* the parameters $\{\lambda, A, \varrho, \eta\}$ in terms of the angles of the exact parametrization in Eq. (1.10) as follows:

$$\lambda \doteq s_{12}, \qquad A\lambda^2 \doteq s_{23}, \qquad A\lambda^3(\varrho - i\eta) \doteq s_{13}e^{-i\delta}. \quad (1.12)$$

The change of variables $\{s_{ij}, \delta\} \to \{\lambda, A, \varrho, \eta\}$ in Eq. (1.10) leads to an exact parametrization of the CKM matrix in terms of the Wolfenstein parameters. Expanding this expression up to $\mathcal{O}(\lambda^5)$ leads to

$$V_{\text{CKM}} = \begin{pmatrix} 1 - \frac{1}{2}\lambda^2 - \frac{1}{8}\lambda^4 & \lambda + \mathcal{O}(\lambda^7) & A\lambda^3(\bar{\varrho} - i\bar{\eta}) \\ -\lambda + \frac{1}{2}A^2\lambda^5[1 - 2(\bar{\varrho} + i\bar{\eta})] & 1 - \frac{1}{2}\lambda^2 - \frac{1}{8}\lambda^4(1 + 4A^2) & A\lambda^2 + \mathcal{O}(\lambda^8) \\ A\lambda^3(1 - \bar{\varrho} - i\bar{\eta}) & -A\lambda^2 + \frac{1}{2}A\lambda^4[1 - 2(\bar{\varrho} + i\bar{\eta})] & 1 - \frac{1}{2}A^2\lambda^4 \end{pmatrix},$$

$$(1.13)$$

where

$$\bar{\varrho} = \varrho\left(1 - \frac{\lambda^2}{2}\right) + \mathcal{O}(\lambda^4), \qquad \bar{\eta} = \eta\left(1 - \frac{\lambda^2}{2}\right) + \mathcal{O}(\lambda^4). \quad (1.14)$$

The advantage of this generalization of the Wolfenstein parametrization is the absence of relevant corrections to $V_{us}$, $V_{cd}$, $V_{ub}$ and $V_{cb}$, and a simple change in $V_{td}$, which facilitate the implementation of experimental constraints.

A key feature of the CKM matrix is that it contains the only complex coupling of the SM Lagrangian. As such, it is responsible for the amount of matter-antimatter asymmetry of the model. More precisely, a non-vanishing value for the phase $\delta$ in Eq. (1.10) or, equivalently, a non-vanishing $\eta$ in Eq. (1.11), implies a violation of the *CP* symmetry, the discrete symmetry obtained combining and charge-conjugation (*C*) which transform particles into anti-particles, and parity $(P)$ [35]. The presence of a single $CP$-violating phase in the SM implies a strong correlation among all $CP$-violating processes: testing such correlations provides a very efficient way to test the model and constrain physics beyond the SM.[7] As we illustrate in the following, an efficient way to make such test is via the so-called CKM *unitarity triangles*.

The unitarity of the CKM matrix implies two type of relations among its elements [35]:

$$\sum_{i=1}^{3} |V_{ij}|^2 = 1, \quad j = 1, 2, 3 \tag{1.15}$$

$$\sum_{i=1}^{3} V_{ji} V_{ki}^* = \sum_{i=1}^{3} V_{ij} V_{ik}^* = 0, \quad j, k = 1, 2, 3, j \neq k. \tag{1.16}$$

Taking each term in Eq. (1.16) as a point in the complex plane, allows us to form six triangles, as shown in Fig. 1.5. All these triangles have and equal area which, not surprisingly, is proportional to $\sin \delta$. It is worth stressing that the relations in Eq. (1.16) are invariant under any phase transformation of the quark fields. Under such transformations the triangles in Fig. 1.5 are

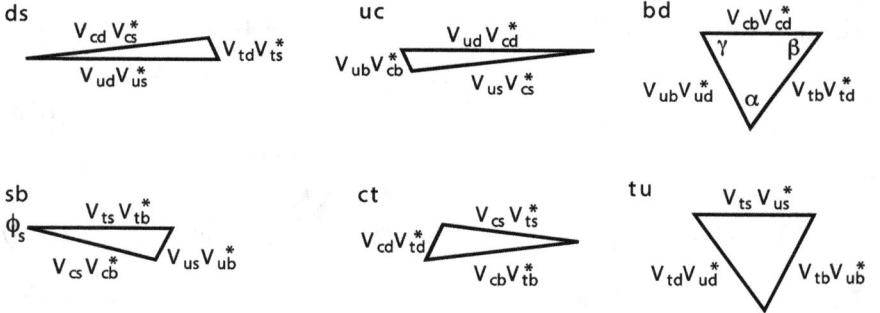

Fig. 1.5: The six CKM triangles. The letters (e.g. **bd**) indicate the row-column multiplication used to generate the triangle (See equation 1.16). Some angles are labeled by their common names. Alternative uses are $\phi_1 = \beta$, $\phi_2 = \alpha$, and $\phi_3 = \gamma$. The triangles are not drawn to scale.

---

[7]In principle, the SM Lagrangian could allow $CP$ violation also in the sector of strong interactions, via the flavor-blind $CP$-odd operator $G\tilde{G}$. However, the coupling of this operator, denoted $\theta_{CP}$, is experimentally bounded to be extremely small ($|\theta_{CP}| < 10^{-10}$) by the limit on the electric dipole moment of the neutron. Once this bound is taken into account, $\theta_{CP}$ has no impact in $B$ physics.

rotated in the complex plane, but angles and sides remain unchanged. Both angles and sides of the unitary triangle are indeed observable quantities which can be measured in suitable experiments.

As we discuss in the next chapter, time-dependent measurements of $CP$-violation in $B$ decays allow us to measure several of the angles in the CKM unitairty triangles, testing this highly constrained structure. This is more difficult for elongated triangles, such as the **bs** triangle, and more easy in the **bd** triangle, where all sizes and angles are of similar size. Looking more closely to the **bd** triangle, the relation defining it is

$$V_{ud}V_{ub}^* + V_{cd}V_{cb}^* + V_{td}V_{tb}^* = 0 \qquad (1.17)$$

or, equivalently,

$$\frac{V_{ud}V_{ub}^*}{V_{cd}V_{cb}^*} + \frac{V_{td}V_{tb}^*}{V_{cd}V_{cb}^*} + 1 = 0 \qquad \leftrightarrow \qquad [\bar{\varrho} + i\bar{\eta}] + [(1 - \bar{\varrho}) - i\bar{\eta}] + 1 = 0.$$

This last form is often referred to as the standard unitarity triangle.

### 1.4.2 *Low-energy effective Lagrangians*

The decays of $B$ mesons are processes which involve at least two different energy scales: the electroweak scale, characterized by the $W$ boson mass, which determines the flavor-changing transition at the quark level, and the $b$ quark mass $(m_b)$, which controls the energy released in the decay. Often also the scale of strong interactions, $\Lambda_{QCD} \sim 300$ MeV $\ll m_b$, related to hadron formation, enters in the evaluation of the decay rates. The presence of these widely separated scales makes the calculation of the decay amplitudes starting from the full SM Lagrangian quite complicated: large logarithms of the type $\log(m_W/m_b)$, or even $\log(m_W/\Lambda_{QCD})$, may appear, leading to a breakdown of ordinary perturbation theory.

This problem can be substantially simplified by integrating out the heavy SM fields ($W$ and $Z$ bosons, as well as the top quark) at the electroweak scale, and constructing an appropriate low-energy effective field theory (EFT) where only the light SM fields appear. The weak effective Lagrangians thus obtained contains local operators of dimension six (and higher), written in terms of light SM fermions, photon and gluon fields, suppressed by inverse powers of the $W$ mass (see Ref. [36] for a more detailed discussion and historical remarks on the construction of effective Lagrangians for weak decays).

To be concrete, let's consider the example of charged-current semileptonic weak interactions. The basic building block in the full SM Lagrangian
is

$$\mathcal{L}_W^{\text{full SM}} = \frac{g}{\sqrt{2}} J_W^\mu(x) W_\mu^+(x) + \text{h.c.}, \qquad (1.18)$$

where

$$J_W^\mu(x) = V_{ij}\, \bar{u}_L^i(x)\gamma^\mu d_L^j(x) + \bar{\nu}_L^j(x)\gamma^\mu \ell_L^j(x) \qquad (1.19)$$

is the weak charged current, whose quark part has already introduced in
Eq. (1.6). Integrating out the $W$ field at the tree level we contract two
vertexes of this type generating the non-local transition amplitude

$$i\mathcal{T} = -i\frac{g^2}{2}\int d^4x D_{\mu\nu}(x, m_W)\, T\left[J_W^\mu(x), J_W^{\nu\dagger}(0)\right], \qquad (1.20)$$

which involves only light fields ($T$ denotes the time-order product of the
two currents). Here $D_{\mu\nu}(x, m_W)$ is the $W$ propagator in coordinate space:
expanding it in inverse powers of $m_W$,

$$D_{\mu\nu}(x, m_W) = \int \frac{d^4q}{(2\pi)^4} e^{-iq\cdot x} \frac{-ig_{\mu\nu} + \mathcal{O}(q_\mu, q_\nu)}{q^2 - m_W^2 + i\varepsilon} = \delta(x)\frac{ig_{\mu\nu}}{m_W^2} + \dots,$$
$$\qquad (1.21)$$

the leading contribution to $\mathcal{T}$ can be interpreted as the tree-level contribution of the following effective local Lagrangian

$$\mathcal{L}_{\text{eff}}^{(0)} = -\frac{4G_F}{\sqrt{2}} g_{\mu\nu} J_W^\mu(x) J_W^{\nu\dagger}(x), \qquad (1.22)$$

where $G_F/\sqrt{2} = g^2/(8m_W^2)$ is the Fermi coupling. If we select in the product
of the two currents the quark current and the electron-neutrino current,

$$\mathcal{L}_{\text{eff}}^{\text{semi-lept}} = -\frac{4G_F}{\sqrt{2}}\, V_{ij}\, \bar{u}_L^i(x)\gamma^\mu d_L^j(x)\, \bar{e}_L(x)\gamma_\mu \nu_L^e(x) + \text{h.c.}, \qquad (1.23)$$

we obtain an effective Lagrangian which provides an excellent description of
semileptonic weak decays to electrons. The neglected terms in the expansion
(1.21) correspond to corrections of $\mathcal{O}(m_B^2/m_W^2)$ to the decay amplitudes.
In principle, these corrections could be taken into account by adding
appropriate dimension-eight operators in the effective Lagrangian. However,
in most cases they are safely negligible.

The case of charged semileptonic decays is particularly simple since we
can ignore QCD effects: the operator in Eq. (1.23) is not renormalized by

strong interactions. The situation is slightly more complicated in the case of non-leptonic or flavor-changing neutral-current processes, where QCD corrections and higher-order weak interactions cannot be neglected, but the basic strategy is the same. First of all we need to identify a complete basis of local operators, that includes also those generated beyond the tree level. In general, given a fixed order in the $1/m_W^2$ expansion of the amplitudes, we need to consider all operators of corresponding dimension (e.g. dimension six at the first order in the $1/m_W^2$ expansion) compatible with the symmetries of the system. Then we must introduce an artificial scale in the problem, the renormalization scale $\mu$, which is needed to regularize QCD (or QED) corrections in the effective theory.

The effective Lagrangian for generic $|\Delta B| = 1$ processes, i.e. processes where one $b$ (or one $\bar{b}$) quark is destroyed, assumes the form

$$\mathcal{L}_{|\Delta B|=1} = -4\frac{G_F}{\sqrt{2}} \sum_i \hat{C}_i(\mu) Q_i, \qquad (1.24)$$

where the sum runs over the complete basis of operators. The effective couplings $\hat{C}_i(\mu)$, known as Wilson coefficients, are dimension-less quantities which depend, in general, on the renormalization scale $\mu$. The dependence from CKM factors in a given effective Hamiltonian is usually quite simple due to flavor symmetries. It is thus convenient to show it explicitly defining CKM-independent Wilson coefficients,

$$\hat{C}_i \to \hat{\lambda}\, C_i, \qquad (1.25)$$

where $\hat{\lambda}$ is an appropriate combination of CKM matrix elements (e.g. $\hat{\lambda} = V_{cb}$, for $b \to ce\bar{\nu}$ decays).

The dependence from the renormalization scale cancels when evaluating the matrix elements of the effective Lagrangian for physical processes, that we can generically indicate as

$$\mathcal{M}(i \to f) = -4\frac{G_F}{\sqrt{2}} \sum_i \hat{\lambda} C_i(\mu) \langle f|Q_i(\mu)|i\rangle. \qquad (1.26)$$

The independence of $\mathcal{M}$ from $\mu$ holds for any initial and final state, including partonic states at high energies. This implies that the $C_i(\mu)$ obey a series of renormalization group equations (RGE), whose structure is completely determined by the anomalous dimensions of the effective operators. These equations can be solved using standard renormalization-group techniques, allowing the resummation of all large logs of the type $\alpha_s(\mu)^{n+m} \log(m_W/\mu)^n$ to all orders in $n$ (working at order $m + 1$ in perturbation theory). The scale $\mu$ acts as a separator of short- and long-distance virtual corrections: short-distance effects are included in the $C_i(\mu)$, whereas long-distance effects are left as explicit degrees of freedom in the effective theory.[8]

In practice, the problem reduces to the following three well-defined and independent steps (see Ref. [36] for an in-depth explanation of each step):

1. the evaluation of the *initial conditions* of the $C_i(\mu)$ at the electroweak scale ($\mu \approx m_W$);
2. the evaluation of the anomalous dimension of the effective operators, and the corresponding *RGE evolution* of the $C_i(\mu)$ from the electroweak scale down to the energy scale of the physical process ($\mu \approx m_B$);
3. the evaluation of the *matrix elements* of the effective Lagrangian for the physical hadronic processes (which involve energy scales from $m_B$ down to $\Lambda_{QCD}$).

The first step is the one where physics beyond the SM may appear: if we assume NP is heavy, it may modify the initial conditions of the Wilson coefficients at the high scale, as we briefly outlined in section 1.4.4. On the other hand, heavy NP cannot directly affect the following two steps. While the RGE evolution and the hadronic matrix elements are not directly related to NP, they may influence the sensitivity to NP of physical observables. In particular, the evaluation of hadronic matrix elements is potentially affected by non-perturbative QCD effects: these are often a large source of theoretical uncertainty which can obscure NP effects. RGE effects do not induce sizable uncertainties since they can be fully handled within perturbative QCD; however, the sizable logs generated by the RGE running may dilute the interesting short-distance information encoded in the high-scale coefficients of selected operators, if those get mixed with operators which are insensitive to NP. As we will discuss in section 1.4.3, only in specific classes of observables these two effects are small and under good theoretical control.

### 1.4.3   *Hadronic matrix elements*

All non-perturbative effects are confined in the hadronic matrix elements of the operators of the effective Lagrangians. As far as the evaluation of the

---

[8]This statement would be correct if the theory were regularized using a dimensional cut-off. It is not fully correct if $\mu$ is the scale appearing in the (often adopted) dimensional-regularization + minimal-subtraction (MS) renormalization scheme.

matrix elements is concerned, we can divide $B$-physics observables in three main categories, as described below.

- *Inclusive decays.* By inclusive decays we mean processes obtained summing over all the hadronic final state with a given quark flavor. For instance $B \to X_c e\bar{\nu}$ stands for the decay of the $B$ meson into a final state that contains an electron, an anti-neutrino, and any combination of hadrons with a $c$ quark and an arbitrary number of quark-anti-quark pairs. The heavy quark symmetry [37–39], the heavy quark expansion (HQE) [40, 41], based on the optical theorem for inclusive decays and the expansion in terms of local operators (operator product expansion) for the transition matrix elements, and finally Soft Collinear Effective Theory (SCET) [42, 43], form a solid theoretical framework to evaluate the hadronic matrix elements for these processes: inclusive hadronic rates are related to those of free $b$ quarks, calculable in perturbation theory, by means of a systematic expansion in inverse powers of $\Lambda_{\text{QCD}}/m_b$. With the assumption of quark-hadron duality, the lowest-order terms in this expansion are the pure partonic rates, and for sufficiently inclusive observables higher-order corrections are usually very small. This technique has been proven to be very successful in the case of charged-current semileptonic decays such as $B \to X_c e\bar{\nu}$, and rare decays such as $B \to X_s\gamma$. However, it has a limited domain of applicability, due to the difficulty of selecting and reconstructing hadronic inclusive states. Its applicability is limited to inclusive processes which can be reconstructed without hard phase space cuts. Thus it cannot be applied to suppressed decays when they require large phase-space cuts to suppress the background, such as $b \to u$ transitions.
- *Single-hadron exclusive decays.* For processes with at most one hadron in the final state, we can often factorise the matrix element into the hadronic matrix element of a quark current, and a non-hadronic term. For instance in $B^- \to e^-\bar{\nu}$ decays, where the $b$ annihilates with the $\bar{u}$ quark to form a virtual $W^-$ that materializes as an $e^-\bar{\nu}$ pair [see Fig. 1.1(c)], the matrix element of the operator in Eq. (1.23) can be decomposed as

$$\langle e^-\bar{\nu}| \bar{u}_L\gamma^\mu b_L\, \bar{e}_L\gamma_\mu \nu_L^e |B^-\rangle = H^\mu \times L_\mu\,, \qquad (1.27)$$

with

$$H^\mu = \langle 0| \bar{u}_L\gamma^\mu b_L |B^-\rangle\,, \qquad L_\mu = \langle e^-\bar{\nu}| \bar{e}_L\gamma_\mu \nu_L^e |0\rangle\,. \qquad (1.28)$$

The hadronic matrix elements of a quark current between single (or no) hadron states are quite simple. For instance, in the case of no hadrons in the final state, we can parameterise it in term of the meson decay constant, $F_{B_q}$, defined by

$$\langle 0|\bar{b}\gamma_\mu\gamma_5 q|B_q(p)\rangle = ip_\mu F_{B_q}. \tag{1.29}$$

For decays with a single hadron in the final states we need to introduce appropriate form factors [see Fig. 1.1(a)]. For instance, if the hadron is a generic pseudoscalar meson ($P$), the matrix element of the quark vector current is expressed in terms of two form factors, $f_{+,o}^{B\to P}(q^2)$, defined by

$$\langle P(k)|\bar{b}\gamma^\mu q|B(p)\rangle = \left[(p+k)^\mu - \frac{m_B^2 - m_P^2}{q^2}q^\mu\right] f_+(q^2)$$
$$+ \frac{m_B^2 - m_P^2}{q^2}q^\mu f_0(q^2) \tag{1.30}$$

where $q^2 = (p-k)^2$. The number of form factors increases with the spin of the hadrons in the initial and final states, since non-vanishing spins allows us to construct more kinematical invariants. The heavy quark effective theory (HQET) [44, 45] introduces relationship between different form factors. For example, in the static approximation, the four form factors describing $B \to D^*\ell\nu$ semileptonic decay can all be expressed in terms of a single unknown function. Corrections to the static approximation can be cast as a power series expansion in $1/m_b$ and $1/m_c$.

Lattice QCD is the best tool to evaluate these non-perturbative quantities from first principles. At present, the form-factors relevant for some "gold-plated" $B$ decays are computed on the lattice with good accuracy (see Ref. [46] and references therein) with a precision that reach the $\sim$2% level in the case of the meson decay constants. An alternative approach is represented by Light Cone Sum Rules (LCSR) (see e.g. [47–50]). In the latter case the estimate of theory errors is less reliable; however, the regime where LCSR is more effective (i.e. for light hadrons in the final state and small momentum transfer) is the one where Lattice QCD has the most severe difficulties. The two approaches are therefore highly complementary.

A key point to stress is that in rare neutral-current decays, such as the processes $B \to K^{(*)}\ell^+\ell^-$ that we will discuss at length in chapter 3.1, the relevant effective Lagrangian contains also four-quark operators. For these operators we cannot perform a simple factorization of the type in Eq. (1.27). As we shall discuss, this complicates significantly the description of these modes introducing irreducible sources of theoretical uncertainty.

- *Multi-hadron final states.* The last class of hadronic matrix elements is the one relevant to multi-hadron final states, such as the two-body non-leptonic decays $B \to \pi\pi$ and $B \to K\pi$, as well as many other processes with more than one hadron in the final state. These are the most difficult ones to be estimated from first principles with high accuracy. A lot of progress in the recent past has been achieved thanks to QCD factorization [51–54] and the SCET [42, 55] approaches, which provide factorization formulae to relate these hadronic matrix elements to two-body hadronic form factors in the large $m_b$ limit. However, it is fair to say that the errors associated to the $\Lambda_{\rm QCD}/m_b$ corrections are still quite large. This subject is quite interesting by itself, but is beyond the scope of this book, where we focus on clean $B$-physics observables for NP studies. To this purpose, the only interesting non-leptonic channels are those where, with suitable ratios, or using precise symmetry relations among hadronic matrix elements, we can eliminate completely all hadronic unknowns. Examples of this type are the $B \to DK$ channels discussed in the next chapter for the extraction of the CKM phase $\gamma$.

### 1.4.4 *Effective Lagrangians for physics beyond the SM*

The procedure of building effective Lagrangians is a perfectly appropriate way to describe the impact of heavy new physics in general terms. The SM Lagrangian becomes the renormalizable part of a more general local Lagrangian which includes an infinite tower of operators with dimension $d > 4$, constructed in terms of SM fields and suppressed by inverse powers of an effective scale $\Lambda$. These operators are the residual effect of having integrated out the new heavy degrees of freedom, whose mass scale is parametrized by the effective scale $\Lambda > m_W$. In the next chapters we will discuss both generic analyses of new physics based on effective Lagrangians, as well as analyses based on explicit extensions of the SM.

When applying the method of effective Lagrangians to describe NP we do not know the nature of the degrees of freedom we are integrating out. We are somehow in the same situation that Fermi faced when proposing his theory of weak decays: we cannot determine, a priori, the values of the effective couplings of the higher-dimensional operators. We can only classify their structure using general symmetry arguments. The advantage of this approach is that it allows us to analyse all the extensions of the SM characterised by heavy degrees of freedom (i.e. physics above the electroweak scale) in terms of a limited number of parameters (the coefficients of the higher-dimensional operators). The drawback is a limitation in establishing correlations between NP effects at low energies and on-shell contributions of the new degrees of freedom at high energies. Moreover, the EFT approach does not allow us to fully reconstruct dynamical aspects of the NP model which can connect the effective couplings involving different sets of fields.

Assuming that a single elementary Higgs field is responsible for the $SU(2)_L \times U(1)_Y \rightarrow U(1)_Q$ spontaneous breaking, as indicated by all available high-energy data, the Lagrangian of the SM considered as an effective field theory can be written as

$$\mathcal{L}_{\text{eff}} = \mathcal{L}_{\text{gauge}}^{\text{SM}} + \mathcal{L}_{\text{Higgs}}^{\text{SM}} + \mathcal{L}_{\text{Yukawa}}^{\text{SM}} + \Delta\mathcal{L}_{d>4}. \tag{1.31}$$

Here $\Delta\mathcal{L}_{d>4}$ denotes the series of higher-dimensional operators invariant under the SM gauge group:

$$\Delta\mathcal{L}_{d>4} = \sum_{d>4} \sum_{n=1}^{N_d} \frac{c_n^{(d)}}{\Lambda^{d-4}} Q_n^{(d)}(\text{SM fields}). \tag{1.32}$$

The Lagrangian in Eq. (1.31) is usually referred to as the SM Effective Field Theory (or SMEFT) Lagrangian.

The expression of the SMEFT Lagrangian is particularly simple if we truncate the expansion at $d = 5$. As shown by Weinberg [56], at this order there is a single electroweak structure, which violates total lepton number and describes non-vanishing (Majorana-type) masses for the left-handed neutrinos when the Higgs field acquires a vacuum expectation value

$$\Delta\mathcal{L}_{d=5} = \frac{1}{\Lambda}(\bar{L}_L^c)^i \lambda_{ij}(L_L)^j H_c^* H_c^\dagger \longrightarrow m_\nu^{ij} = \lambda_{ij}\frac{v^2}{2\Lambda}. \tag{1.33}$$

A much richer structure arises at $d = 6$, where a plethora of new operators appear. In particular, at this order there are several four-fermion operators which describe in general terms the effects of heavy new physics in $b$-quark decays. The first complete classification of these operators, with special focus on the possible violation of flavor in extensions of the SM, dates back to the pioneering work by Buchmuller and Wyler in 1985 [57]. It took about 25 years before a systematic analysis of all the $d = 6$ operators, aimed at obtaining a non-redundant basis, was performed [58]. The basis identified in Ref. [58] is usually referred to as the Warsaw basis of the SMEFT. Using such basis, the complete one-loop renormalization-group structure of the effective theory (up to $d = 6$ terms) has been explicitly computed in Ref. [59–61].

### 1.4.5 *Accidental and approximate symmetries*

If NP appears at the TeV scale, as we expect from the stabilization of the mechanism of electroweak symmetry breaking, the scale $\Lambda$ characterising the suppression of the higher dimensional operators in the SMEFT should not exceed a few TeV. More precisely, we expect such a low effective scale at least for the operators involving the Higgs fields, and fields strongly coupled to the Higgs sector, such as third-generation fermions. Moreover, we expect $c_i^{(d)} = \mathcal{O}(1)$ for all the operators which are not forbidden (or suppressed) by specific symmetry structures in the underlying theory. These two requirements are apparently in strong conflict in several dimension-six operators contributing to flavor-changing processes. These operators must have very small coefficients to be compatible with present data, if the overall scale of the effective theory is set to a few TeV. We will quantify this statement more precisely in the next chapter when discussing $B$–$\bar{B}$ mixing.

An even stronger conflict is found for operators violating baryon- or lepton-number, such as operators contributing to proton decay, or the dimension-five term in Eq. (1.33) describing neutrino masses (which is non-zero but has a very tiny coupling). However, baryon number ($B$) and lepton number ($L$) are exact global symmetries of the $d = 4$ part of the Lagrangian. They are denoted *accidental symmetries*, since these global symmetries arise automatically in the lowest-dimensional operators of the effective theory as indirect consequences of the field content and the gauge symmetry. The strong bounds on the $B$-violating terms (from proton stability), and the tiny coefficient of the $L$-violating term in Eq. (1.33), indicate that such symmetries remain unbroken up to very heavy scales. This is not in contradiction with the expectation of a few TeV scale for the operators which preserve $B$ and $L$ in the rest of the SMEFT, since the symmetry-preserving sector cannot induce violations of the symmetries. The accidental symmetries allow us to conceive a multi-scale structure in the underlying theory: a low effective scale in the symmetry-preserving sector of the EFT does not prevent a possible high scale (i.e. a strong suppression) for the symmetry-breaking terms.

The case of flavor-violating terms (preserving $B$ and $L$) is apparently different, since there is no exact flavor symmetry in the low-energy theory: flavor quantum numbers are not conserved in the SM Lagrangian. However, as discussed at the beginning of this chapter, the SM Lagrangian has a large flavor symmetry in the gauge sector, which is broken in a rather peculiar way by the Yukawa couplings. Since most of the breaking terms are small, in this case we can speak of *approximate accidental symmetries*. In practice, there is no real difference between exact and approximate accidental symmetries: in both cases we can conceive a multi-scale structure in order to preserve a low effective scale in the symmetry-preserving sector of the theory. More explicitly, in order to keep a relatively low (few TeV) effective scale in the SMEFT, we need to assume that at such energy scale not only $B$ and $L$, but also the tightly constrained accidental flavor symmetries, remain valid, or are broken only by small symmetry-breaking terms. Specific examples on how this can be achieved are discussed in appendix B. The technical implementation is obtained imposing a set of symmetries and symmetry breaking terms, known as *spurions*, in the SMEFT. This technical implementation can be viewed as a pragmatic way to deal with a possible underlying multi-scale structure.

On the other hand, searching for possible violations of the approximate accidental flavor symmetries of the SM Lagrangian represents a very powerful tool to search for physics beyond the SM. That's exactly what makes the study of $B$ physics particularly interesting. If the high-scale NP does not respect such symmetries, we should be able to observe violations also at low energies, with enough precision. The key point is that observing the violation of a symmetry, even if tiny, is usually simpler than looking for deviations from the theory predictions in SM-allowed processes (i.e. processes which do not violate any symmetry of the low-energy sector of the effective theory).

## Lepton Flavor Universality

Among the approximate flavor symmetries, one that plays a central role in this book, in view of recent results in $B$ physics, is *Lepton Flavor Universality* (LFU). We can define LFU as the permutation symmetry among the three different lepton species. This is not an exact symmetry of the SM Lagrangian, but it becomes such in the limit where we neglect the three lepton Yukawa couplings

$$y_e = \frac{\sqrt{2}m_e}{v} \approx 3 \times 10^{-6}, \quad y_\mu = \frac{\sqrt{2}m_\mu}{v} \approx 0.6 \times 10^{-3},$$

$$y_\tau = \frac{\sqrt{2}m_\tau}{v} \approx 10^{-2}. \tag{1.34}$$

As can be seen from the figures above, the three lepton Yukawa couplings are all small compared to the SM gauge couplings:

$$y_{e,\mu,\tau} \ll g_{1,2,3}. \tag{1.35}$$

This is why it is a good approximation to neglect them when evaluating transition amplitudes. More precisely, within the SM is a good numerical approximation to neglect the lepton-Yukawa *forces* compared to all the other interactions, while retaining only their ground-state effect, i.e. the non-vanishing masses for the charged leptons.

Summarizing, within the SM the only sizable violations of LFU are kinematical effects due non-vanishing lepton masses. However, there is nothing fundamental in LFU: it is an approximate accidental low-energy symmetry of the SM Lagrangian, as many others in the flavor sector. Recent data discussed in chapters 3 and 4 seem to indicate that LFU is violated not only by the lepton Yukawa couplings. If confirmed, these new sources of LFU violations provide a very interesting and non-trivial clue on high-scale dynamics.

# Chapter 2

# Traditional New Physics searches

As discussed in the previous chapter, the prerequisite for most NP searches is a precise determination of the SM couplings. Among them, a key role is played by the elements of the CKM matrix, whose study is the main subject of this chapter. One of the appealing feature of $B$ physics is that it allows the determination of several CKM elements in different ways, taking advantage of different amplitudes. These measurements overconstrain the Standard Model, thus providing a powerful check of its ability of describing $b$ decays and set bounds on New Physics models.

This chapter is organised in the following way: section 2.1 is devoted to the determination of the CKM elements $|V_{cb}|$ and $|V_{ub}|$ via tree-level semileptonic decays involving light leptons. The $CP$-violating phase $\gamma$ discussed in section 2.2 is another tree-level quantity expected to provide a theoretically-clean check of SM predictions. In section 2.3 we will introduce the phenomena of $B$–$\bar{B}$ mixing and CP violation. Lastly, in section 2.4 we discuss how these measurements are combined in global fits to assess their overall SM compatibility, and the constraints that they provide for a large class of NP models.

## 2.1 Determining the size of the CKM elements

### 2.1.1 *Measurements of $|V_{cb}|$*

Semileptonic decays, illustrated in Fig. 2.1, are used to determine the parameter $|V_{cb}|$. They are tree-level diagrams, and thus they are expected to be well described by the Standard Model, although lepton flavor universality violation in tauonic versus muonic decays may alter this view, if it becomes well established (see section 3.4). The main challenge is the calculation of

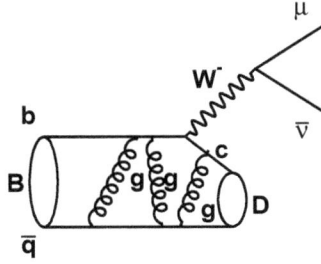

Fig. 2.1: Feynman diagram describing $B$-hadron semileptonic decay to a charmed hadron, used to measure the quark mixing parameter $|V_{cb}|$. $D$ indicates one possible charmed hadron and $g$ gluons.

the hadronic current describing the $B$ to $D^{(*)}$ transition, as the energy scale of the process mandates its evaluation with non-perturbative QCD calculations.

There are two distinct approaches to extract these quark mixing parameters: the first, known as the *exclusive* approach, encompasses measurements where the hadron accompanying the lepton in the final state is reconstructed; the second, known as the *inclusive* approach, involves the study of inclusive properties of the final state, such as the lepton momentum spectrum or $q^2$ distribution, where $q^2$ is the four-momentum squared transfer between the $B$ and charmed-hadron. These two methods have different experimental systematic and theoretical uncertainties.

We describe first the exclusive approach, involving the study of a specific semileptonic decay, using the extensively studied $B^0 \to D^{*-}\mu^+\nu_\mu$ decay. In this case the differential decay width can be expressed in terms of $q^2$, two helicity angles describing the orientation of the charged lepton in the dilepton rest frame, $\theta_\ell$, and the orientation of the $D^0$ in the $D^{*-}$ rest frame, $\theta_V$, and $\chi$, the angle between the two decay planes in the $B^0$ rest frame (see Fig. 2.2).

For transitions among heavy quarks ($b \to c$), the variable $q^2$ is often replaced by the four-velocity transfer

$$w = \frac{\left(m^2_{B^0} + m^2_{D^{*-}} - q^2\right)}{(2M_{B^0}m_{D^{*-}})}. \tag{2.1}$$

The use of $w$ is motivated by the heavy quark symmetry discussed in the introduction, which predicts that semileptonic transitions between a $b$ quark and a $c$ quark at $w = 1$ have a form factor equal to unity modulo some small and calculable QCD corrections [37, 38].

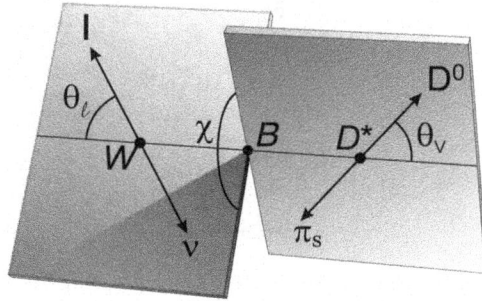

Fig. 2.2: Definition of the angles $\theta$, $\theta_V$, and $\chi$ used to describe the decay $B \to D^* \ell \bar{\nu}$.

The differential decay width for $B^0 \to D^{*-}\mu^+\nu_\mu$ is governed by four form factors defined as

$$\langle D^* | \bar{c}\gamma^\mu b | \bar{B} \rangle = i\sqrt{m_B m_{D^*}}\, h_V\, \varepsilon^{\mu\nu\alpha\beta}\, \epsilon^*_\nu v'_\alpha v_\beta,$$

$$\langle D^* | \bar{c}\gamma^\mu\gamma^5 b | \bar{B} \rangle = \sqrt{m_B m_{D^*}}\, \big[ h_{A_1}(w+1)\epsilon^{*\mu}$$

$$- h_{A_2}(\epsilon^* \cdot v)v^\mu - h_{A_3}(\epsilon^* \cdot v)v'^\mu \big], \qquad (2.2)$$

where $v$ is the four-velocity of the $B$, $v'$ is that of the $D^*$, and $\epsilon^*$ indicates the three possible $D^*$ polarization states. The $h_i$ indicate the four form factors.

Early on, in the absence of theory tools to calculate these form factors, phenomenological models, for example Refs. [62, 63], were developed to derive the shapes and normalizations from a combination of different anzatz's and quark model calculations. A key step towards the development of a precision method to measure $|V_{cb}|$ with the exclusive approach was achieved through the discovery of the highly symmetric structure of transitions involving quarks with mass $m_Q \gg \Lambda_{QCD}$, which led to the formulation of HQET [37, 38]. For $m_Q \gg \Lambda_{QCD}$, the heavy quark behaves as an almost static source of the chromodynamic field. As a result, in first approximation the dynamics of the light degrees of freedom inside the hadron is independent of both the heavy-quark flavor and its spin. This observation leads to a highly simplified lowest order expression of the exclusive $b \to c$ form factors with a normalization dictated by the theory. In particular, all the four form factors are related to an unknown function $\xi(w)$, known as the Isgur-Wise (IW) function [37] with perturbative corrections parameterized in powers of $\alpha_s$ and non-perturbative corrections that can be organized as an expansion in powers of $1/m_b$. In $b \to c$ transitions coefficients in powers of $1/m_c$ also need to be considered. In some cases, such as $B \to D^*$ transitions, the symmetry dictates the vanishing of lowest order corrections.

The differential decay width can then be expressed as [12]

$$\frac{d\Gamma}{dw}\left(B^0 \to D^{*-}\mu^+\nu_\mu\right) = \frac{G_F^2 m_B^5}{48\pi^3}|V_{cb}|^2(w^2-1)^{1/2}P(w)\,(\eta_{EW}\mathcal{F}(w))^2, \tag{2.3}$$

where $P(w)$ is a phase space factor, $\eta_{EW} = 1.0066 \pm 0.0050$ [64, 65] accounts for electroweak corrections, and $P(w)\mathcal{F}(w)$ is given by

$$P(w)|\mathcal{F}(w)|^2 = |h_{A_1}(w)|^2 \left\{2\frac{r^2-2rw+1}{(1-r)^2}\left[1+\frac{w-1}{w+1}R_1^2(w)\right]\right.$$
$$\left.+\left[1+\frac{w-1}{1-r}(1-R_2(w))\right]^2\right\}, \tag{2.4}$$

where $r = m_D^*/m_B$ and $R_1(w)$ and $R_2(w)$ are the form factor ratios

$$R_1(w) = \frac{h_V(w)}{h_{A_1}(w)}, \quad R_2(w) = \frac{h_{A_3}(w) + (m_{D^*}/m_B)h_{A_2}(w)}{h_{A_1}(w)}. \tag{2.5}$$

The form factor normalization $\mathcal{F}(1)$ is obtained from LQCD calculations averaged in Ref. [46] with the most recent results and is found to be $\mathcal{F}(1) = 0.904 \pm 0.013$. A commonly used parameterization of the form factor $h_{A_1}(w)$, proposed in Ref. [66], depends only upon the parameter $\rho^2$, via the expression

$$h_{A_1}(w) = h_{A_1}(1)\left[1 - 8\rho^2 z + (53\rho^2 - 15)z^2 + (231\rho^2 - 91)z^3\right], \tag{2.6}$$

where the variable $z$ is related to $w$ via

$$z = \frac{(\sqrt{w+1} - \sqrt{2})}{(\sqrt{w+1} + \sqrt{2})}. \tag{2.7}$$

Alternative expressions for the relevant form factors, with less stringent HQET constraints [67], are used in other extractions of $|V_{cb}|$ from $B \to D^*\ell\nu$.

The most commonly used method to implement non-perturbative hadronic matrix element calculations is lattice QCD (LQCD). After decades of intensive work on the algorithms and computational tools, several LQCD calculations have achieved remarkable accuracy, of the order of 1%, at least for some "golden" final states, in specific regions of phase-space [68].

Generally, LQCD calculations are more precise at the so called zero-recoil limit, where the invariant mass of the lepton-neutrino pair is maximum. In fact, most of the original unquenched LQCD calculations predicted the form factor normalization only at zero-recoil, however more recent calculations predict form factor shapes, at least in a limited $q^2$ range [69]. HQET provides useful constraints and a prediction for the limit of the theory for very high $m_Q$ that are incorporated in some calculations (see for example Ref. [70]). Most of the theoretical work has focused on semileptonic decays of the $B^+$ or $B^0$ mesons, as the majority of the data were provided by the $e^+e^-$ $b$-factories. LHCb has changed the situation and made it possible to extract $|V_{cb}|$ from $B_s^0$ decays [71]. A measurement of $|V_{cb}|$ from $\Lambda_b^0 \to \Lambda_c^+ \mu^- \bar{\nu}_\mu$ is imminent, following a precise determination of the shape of the differential decay width $d\Gamma/dq^2$ [72]. Recently form factor calculations at non-zero recoil for the decay $B \to D^{*-}\mu^+\nu_\mu$ have been implemented [73], and the paper also includes a joint fit to experimental data. This determination is consistent with previous results and confirms the discrepancy with the inclusive $|V_{cb}|$ evaluation discussed below.

Table 2.1 summarizes the $|V_{cb}|$ exclusive determinations. The averages from Ref. [74] include early $B \to D^*\ell\nu$ measurements, which used a drastically simplified version of HQET, relating all the relevant form factors via the IW function. Their fit included only $|V_{cb}|$, proportional to the intercept at zero-recoil ($w = 1$), and $\rho^2$, the slope of the IW function at $w = 1$. Recent determinations involve more shape parameters in the fit, equivalent to the considerations of effect of subleading IW functions.

The heavy quark expansion is the only method so far available to evaluate the hadronic current in inclusive processes with systematically improvable theoretical uncertainties. It combines an operator product expansion for inclusive rates with HQET. The hadronic tensor $W^{\mu\nu}$ is evaluated adding the contributions of all the possible final states $X_c$, using

Table 2.1: Current data on $|V_{cb}|$ determinations from exclusive $B$ semileptonic decays.

| Decay | $|V_{cb}|$ $[10^{-3}]$ | Fit Model | Reference |
|---|---|---|---|
| $B \to D^*\ell^+\nu_\ell$ | $38.76 \pm 0.42_{\text{exp}} \pm 0.55_{\text{th}}$ | CLN [66] | [74] |
| $B \to D^*\ell^+\nu_\ell$ | $38.40 \pm 0.34_{\text{exp}} \pm 0.66_{\text{th}}$ | Lattice-Data joint BGL fit | [73] |
| $B \to D\ell^+\nu_\ell$ | $39.58 \pm 0.94_{\text{exp}} \pm 0.37_{\text{th}}$ | CLN [66] | [74] |
| $B_s^0 \to D_s^{(*)-}\mu^+\nu_\mu$ | $41.4 \pm 0.6 \pm 0.9 \pm 1.2$ | CLN [66] | [71] |
| $B_s^0 \to D_s^{(*)-}\mu^+\nu_\mu$ | $42.3 \pm 0.8 \pm 0.9 \pm 1.2$ | BGL [75] | [71] |
| Average | $38.90 \pm 0.53$ | | [76] |

the Operator Product Expansion (OPE) similar to the one used in deep inelastic scattering [77]. Then, in the context of heavy quark decays, the OPE is an expansion in powers of $1/m_Q$. This method allows the description of the semileptonic width $\Gamma_{sl}$ in terms of a few parameters. In the kinetic scheme, described in Ref. [78], $\Gamma_{sl}$ is parameterized as

$$
\Gamma_{sl} = \Gamma_0 f(\rho) \left[ 1 + a_1 a_s + a_2 a_s^2 + a_3 a_s^3 - \left( \frac{1}{2} - p_1 a_s \right) \frac{\mu_\pi^2}{m_b^2} \right.
$$
$$
\left. + (g_0 + g_1 a_s) \frac{\mu_G^2(m_b)}{m_b^2} + d_0 \frac{\rho_D^3}{m_b^3} - g_0 \frac{\rho_{LS}^3}{m_b^3} + \cdots \right] \qquad (2.8)
$$

where $\Gamma_0 = A_{ew} |V_{cb}^2| G_F^2 \left[ m_b^{kin}(\mu) \right]^5 / 192\pi^3$, $f(\rho) = 1 - 8\rho + 8\rho^3 - \rho^4 - 12\rho^2 \ln \rho$, $a_s = \alpha_s^{(4)}(\mu_b)/\pi$ is the strong coupling in the $\overline{\text{MS}}$ scheme with 4 active quark flavours, $\rho = \left[ \overline{m}_c(\mu_c)/m_b^{kin}(\mu) \right]^2$ is the squared ratio of the $\overline{\text{MS}}$ charm mass at the scale $\mu_c$, $\overline{m}_c(\mu_c)$, and of $m_b^{kin}(\mu)$, the $b$ quark kinetic mass with a cutoff $\mu \sim 1\,\text{GeV}$; $A_{ew} \simeq 1.014$ is the leading electroweak correction. The parameters $\mu_\pi^2, \rho_D^3$, etc. are nonperturbative expectation values of local operators in the $B$ meson defined in the kinetic scheme with cutoff $\mu$. The non-perturbative parameters in the HQE are obtained through a global fit implemented with a methodology that has been proposed in Ref. [79]. They use all the measured semileptonic moments and the constraint $\overline{m}_c = 0.986 \pm 0.011$ [80]. These masses are defined in the kinetic scheme [77], which avoids potential problems with infrared divergencies known as renormalons [81]. Moments of the hadronic mass spectrum and of the lepton spectrum with different cuts are used. Measurements of a specific moment with different cuts are highly correlated. Table 2.2 shows the most recent fit [82] and the correlation coefficients between different parameters.

Decades of work have refined the theoretical inputs and experimental data on the inclusive and exclusive determinations of $|V_{cb}|$, but a difference of about $3\sigma$ between the two methods is not yet understood.

### 2.1.2 *Measurements of $|V_{ub}|$*

The inclusive and exclusive approaches are used also to evaluate the parameter $|V_{ub}|$. In this case the quark in the final state is light, thus HQET does not apply directly, although it provides some constraints.

Lattice-QCD calculations of the $B \to \pi$ form factors provide the currently most precise method to derive $|V_{ub}|$ with the exclusive approach. Joint fits of lattice "data points" and measurements have been performed to

Table 2.2: Fit result in the kinetic scheme [82], in their default scenario ($\mu_c = 2\,\text{GeV}, \mu_b = m_b^{kin}/2$). All parameters are expressed in GeV at the appropriate power. The first and second rows give the central values and their uncertainties, the correlation matrix follows. The $\chi^2_{min}/dof$ is 0.47.

| $m_b^{kin}$ | $\overline{m}_c(2\text{GeV})$ | $\mu_\pi^2$ | $\rho_D^3$ | $\mu_G^2(m_b)$ | $\rho_{LS}^3$ | $\text{BR}_{c\ell\nu}$ | $10^3|V_{cb}|$ |
|---|---|---|---|---|---|---|---|
| 4.573 | 1.092 | 0.477 | 0.185 | 0.306 | −0.130 | 10.66 | 42.16 |
| 0.012 | 0.008 | 0.056 | 0.031 | 0.050 | 0.092 | 0.15 | 0.51 |
| 1 | 0.307 | −0.141 | 0.047 | 0.612 | −0.196 | −0.064 | −0.420 |
|  | 1 | 0.018 | −0.010 | −0.162 | 0.048 | 0.028 | 0.061 |
|  |  | 1 | 0.735 | −0.054 | 0.067 | 0.172 | 0.429 |
|  |  |  | 1 | −0.157 | −0.149 | 0.091 | 0.299 |
|  |  |  |  | 1 | 0.001 | 0.013 | −0.225 |
|  |  |  |  |  | 1 | −0.033 | −0.005 |
|  |  |  |  |  |  | 1 | 0.684 |
|  |  |  |  |  |  |  | 1 |

the $d\Gamma/dq^2$ distribution in order to reduce the overall uncertainty in $|V_{ub}|$. The lattice data is extrapolated to the continuum over the full $q^2$ range. The resulting value is [74]

$$V_{ub}[B \to \pi\ell\bar{\nu}_\ell] = (3.70 \pm 0.10 \pm 0.12) \times 10^{-3}. \qquad (2.9)$$

In the inclusive approach the theoretical description is based on HQE, which predicts the total rate with uncertainties below 5% [83, 84]. However, in order to suppress the background from the dominant $B \to X_c\ell\bar{\nu}_\ell$ decays, the measured phase space for these decays is reduced because of kinematic selections that restrict the analysis to a region where this background is highly suppressed. These partial decay rates cannot be calculated with HQE: the restricted phase space requires the introduction of non-perturbative distribution functions, known as *shape functions* whose moments can be related to known HQE parameters [85, 86]. Various efforts have been dedicated to the identification of a kinematic region where the background from $B \to X_c\ell\bar{\nu}_\ell$ decays is suppressed and the the partial decay width can be predicted reliably [87–91]. Recently the Belle collaboration published a new determination of $|V_{ub}|_{incl}$ with the GGOU method [92] with an alternative analysis method that allows to probe a bigger phase space region. Their result leads to a drop in the average by 3.7%. Table 2.3 summarizes the five different average value which are obtained with different approaches. Reference [74] chooses one of these methods (GGOU) as

Table 2.3: Measurements of $|V_{ub}|$ from inclusive semileptonic decays and their average based on the methods described in the text. The average values are taken from the 2021 update of the HFLAV averages obtained with the method described in Ref. [74]. The uncertainties quoted are experimental and theoretical respectively. Finally, the $\chi^2/\text{dof}$ reflects the consistency between measurements.

| Method | $|V_{ub}|\,[\times 10^{-3}]$ | $\chi^2/\text{dof}$ |
|---|---|---|
| BLNP [87] | $4.28 \pm 0.13^{+0.20}_{-0.21}$ | 16.1/11 |
| GGOU [88] | $4.19 \pm 0.11^{+0.11}_{-0.12}$ | 15.1/10 |
| DGE [93] | $3.92 \pm 0.10^{+0.11}_{-0.12}$ | 21.4/11 |
| ADFR [90] | $3.92 \pm 0.10^{+0.18}_{-0.12}$ | 30.2/10 |
| BLL [91] | $4.62 \pm 0.20 \pm 0.29$ | 1.4/2 |

a reference value, and quotes

$$|V_{ub}|_{\text{incl}} = [4.19 \pm 0.12^{+0.11}_{-0.12})] \times 10^{-3} (\text{HFLAV}). \qquad (2.10)$$

This result includes the most recent Belle result.

The Particle Data Group [12] uses a more conservative approach, chooses three methods, BLNP, GGOU, and DGE, and performs an arithmetic average of the corresponding values of $|V_{ub}|$ adding an additional uncertainty equal to half of the spread between their values, to account for the fact that the model dependence in the partial branching fractions is sensitive to how the models compare in the restricted phase space with good signal-to-background, not by how they compare when integrated over the full kinematic range used in the fit. If we apply their method [12] to the most recent HFLAV averages, we obtain

$$|V_{ub}|_{\text{incl}} = [4.19 \pm 0.12^{+0.13}_{-0.11} \pm 0.18] \times 10^{-3} (\text{PDG}). \qquad (2.11)$$

These numbers illustrate how difficult it is to obtain a precise value of $|V_{ub}|_{\text{incl}}$ and the necessity to state clearly the value and uncertainties assumed in the global CKM fits described below, especially in the fits relying only on tree-level observables.

The relative uncertainties in the quoted values for the inclusive and exclusive determinations of $|V_{ub}|$ are larger than for $|V_{cb}|$, but also in this case a discrepancy between 1 and $2\sigma$ is as yet unaccounted for.

An interesting measurement, exploiting the large number of $b$-baryons produced at the LHC, determines the ratio $|V_{ub}/V_{cb}|$ from the study of the exclusive decays $\Lambda_b^0 \to p\mu^-\overline{\nu}_\mu$ and $\Lambda_b^0 \to \Lambda_c^+ \mu^- \overline{\nu}_\mu$ [94]. In this case $|V_{ub}/V_{cb}|$ is derived from the expression

$$\frac{\mathcal{B}(\Lambda_b^0 \to p\mu^-\overline{\nu}_\mu)\,|_{q^2>14\,\mathrm{GeV}^2}}{\mathcal{B}(\Lambda_b^0 \to \Lambda_c^+\mu^-\overline{\nu}_\mu)\,|_{q^2>10\,\mathrm{GeV}^2}}\, = \frac{|V_{ub}|^2}{|V_{cb}|^2} \times R_{FF} \qquad (2.12)$$

where $R_{FF} = (0.68 \pm 0.04 \pm 0.,08) \times 10^{-2}$ is the theory normalization [95]. This gives

$$\left|\frac{V_{ub}}{V_{cb}}\right| = 0.083 \pm 0.004 \pm 0.004, \qquad (2.13)$$

where the first uncertainty accounts for experimental effects and the second reflect uncertainties in the theoretical calculation.

Figure 2.3 summarizes the exclusive constraints on the quark mixing parameters $|V_{cb}|$ and $|V_{ub}|$ [74] The point represents the exclusive and inclusive $|V_{cb}|$ values determined with the kinetic scheme and $|V_{ub}|$ obtained with the GGOU method. The uncertainty on $|V_{ub}|$ inclusive does not reflect the spread in central values from alternative determinations.

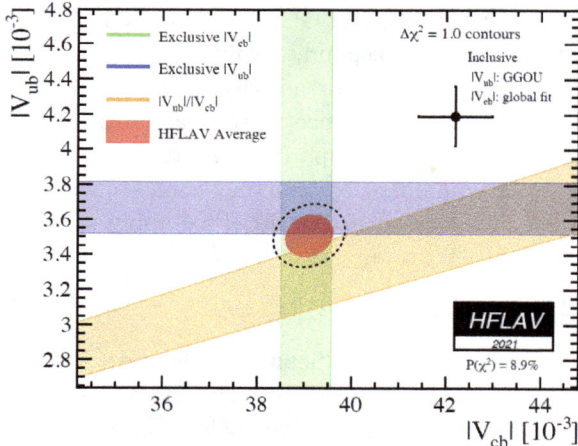

Fig. 2.3: Summary of current knowledge on the parameters $|V_{cb}|$ and $|V_{ub}|$ [74]. Note that the inclusive $|V_{ub}|$ determination quoted is for a specific approach, and does not incorporate the information from the other determinations summarized in Table 2.3.

### 2.1.3 *The composition of the inclusive semileptonics bottom to charm branching fractions*

There are several reasons why a complete mapping of the inclusive semileptonic width of the $b$-hadrons onto exclusive channels is very important. For example, the discrepancies between inclusive and exclusive determinations of $|V_{ub}|$ and $|V_{cb}|$ can be affected by this imperfect knowledge, and the reliability of the Monte Carlo modeling of these decays influences the efficiency evaluation, which is a necessary ingredient to precise determinations of these CKM parameters. Moreover, a full understanding of the Cabibbo-favored decays would make it easier to control the dominant background in charmless semileptonic $B$ hadron decays. The manifestations of lepton flavor universality violations discussed in the next chapter may also be affected by the incomplete understanding of the semileptonic width.

To determine the inclusive branching fractions of the $B^0$ and $B^-$, we use the averages from Ref. [30], which relies on measurements of the lepton spectrum from CLEO [96], BaBar [97], and Belle [98]. The total semileptonic width for $B^+$ and $B^0$ are known to be equal [99], and therefore that the branching fractions differ only by the lifetime difference of $B^+$ and $B^0$. The results of the average [30] are

$$\mathcal{B}(B^- \to X\mu\nu) = (11.09 \pm 0.20)\%$$
$$\mathcal{B}(\overline{B}^0 \to X\mu\nu) = (10.31 \pm 0.19)\%. \tag{2.14}$$

Table 2.4 summarizes the measured branching fractions of semileptonic $B^+$ decays with only a single charm hadron in the final state. Constraints from measured $B^0$ decays into $D^{**}$ mesons are included in this estimate. The unmeasured charm-hadron channels involving a $\pi^0$ in the final state are accounted for using isospin symmetry. It can be seen that the sum of the known components of the inclusive semileptonic width sum up to a number that has a gap of $(1.31 \pm 0.32)\%$ with the known total branching fraction. An estimate of the gap between semi-inclusive measurements and the inclusive branching fraction with a model dependent estimate of the unseen modes involving one or two $\pi^0$ [100] give a similar gap of $(1.07 \pm 0.36)\%$, thus we can conclude that about of 10% of the semileptonic widths of the $B^+$ mesons are not accounted for by measured exclusive modes. This limited knowledge leads to unknown uncertainties in comparing inclusive and exclusive determinations of the CKM parameters and perhaps has implications also for the anomalies in tauonic semileptonic decays discussed

Table 2.4: Summary of the measured and isospin related branching fractions of $B^+$ to charmed final states. Isospin is used to account for decays such as $\bar{D}_0^{*0} \to \bar{D}^0 \pi^0$ from the measured $\bar{D}_0^{*0} \to D^+ \pi^-$.

| Mode | Branching Fraction (%) | Reference |
|------|------------------------|-----------|
| $B^+ \to \bar{D}^0 \mu^+ \nu_\mu$ | $2.29 \pm 0.08$ | [12] |
| $B^+ \to \bar{D}^{*0} \mu^+ \nu_\mu$ | $5.66 \pm 0.22$ | [12] |
| $B^+ \to \bar{D}_0^{*0} \mu^+ \nu_\mu$ | $0.38 \pm 0.08$ | [12] |
| $B^+ \to \bar{D}_1^{\prime 0} \mu^+ \nu_\mu$ | $0.41 \pm 0.09$ | [12] |
| $B^+ \to \bar{D}_1^0 \mu^+ \nu_\mu$ | $0.45 \pm 0.03$ | [12] |
| $B^+ \to \bar{D}_2^{*0} \mu^+ \nu_\mu$ | $0.39 \pm 0.04$ | [12] |
| $B^+ \to \bar{D}_1^{\prime 0}(\bar{D}_1^{\prime 0} \to \bar{D}^0 \pi^+ \pi^-)\mu^+ \nu_\mu{}^{(1)}$ | $0.19 \pm 0.01$ | [12] |
| $B^+ \to X_c \mu^+ \nu_\mu$ | $11.09 \pm 0.28$ | [30] |
| Gap [inclusive versus exclusive] | $1.32 \pm 0.33$ | |
| Gap [inclusive versus semi-inclusive] | $1.07 \pm 0.36$ | [100] |

$^{(1)}$This contribution has been evaluated with the method used in Ref. [101].

in the next chapter. Similar arguments apply to charmless semileptonic $B$ decays, for which only a few exclusive channels have been studied so far.

## 2.2   The measurement of $\gamma$

The angle $\gamma$ in the standard CKM triangle is defined as

$$\gamma \equiv \arg\left[-\frac{V_{ud}V_{ub}^{*}}{V_{cd}V_{cb}^{*}}\right]. \qquad (2.15)$$

This equation shows that $\gamma$ does not depend on CKM elements involving the top quark, and thus it can be measured through interference effects between different tree-level diagrams. In a wide class of SM extensions its determination is expected to be insensitive to physics beyond the SM, which affects only loop-induced processes [102]. Thus, a comparison between the measured value of $\gamma$ and the value inferred from the global CKM fits discussed in section 2.3.4 (which depend on loop-induced amplitudes via $B$-$\bar{B}$ mixing) is an interesting SM test. Several approaches to the experimental determination of $\gamma$ have been pursued, mostly at the Belle, BaBar and LHCb experiments. They exploit the interference between $b \to c\bar{u}s$ and $b \to u\bar{c}s$ diagrams in processes such as $B \to DK$ and related modes. Figure 2.4 shows the two interfering amplitudes and the mechanism through which the angle $\gamma$ affects them.

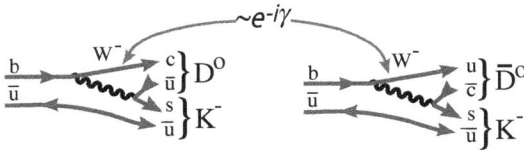

Fig. 2.4: Interference between $b \to c\bar{u}s$ and $b \to u\bar{c}s$ diagrams in charged $B \to DK$ decay, illustrating its relationship to $\gamma$, which is the weak phase between $V_{ub}$ and $V_{cb}$.

The methods proposed to measure $\gamma$ choose different neutral charm meson decays, where the final states can come frome either $D^0$ or $\overline{D}^0$ initial states. In all cases the interference between strong and weak phases is exploited, where the weak phases change sign from particle to anti-particle while the strong phases do not. The first approach (GLW) [103, 104] focuses on $D^0$ decays to CP eigenstates, such as $D^0 \to K^+ K^-$ or $D^0 \to \pi^+ \pi^-$. The second approach (ADS) [105] considers the interference between doubly-Cabibbo suppressed [105] and color-suppressed final states. The last approach in charged $B$ decays considers $B^+$ transitions to $D^0$ mesons decaying into self-conjugate three-body final states (e.g. $D^0 \to K^0 \pi^+ \pi^-$) and a $K^+$ (BPGGSZ) [106–108]. A complementary method is based on the measurement of time-dependent CP parameters in the interference between mixing and decay amplitudes [109, 110] in $B^0$ and $B^0_s$ decays, they will be referred as TD. These measurements are challenging, as they involved time-dependent fits to decay spectra of flavor-tagged $B$ mesons. However their precision will improve with the large data sets expected for the next phase of flavor experiments, and their systematic uncertainties are complementary to the ones in $B^+$ determinations.

In the case of a charged $B \to DK$ with the $D$ meson decaying into a final state $f$ accessible both to the $D^0$ and $\overline{D}^0$, the charge asymmetry $A \equiv (\Gamma_- - \Gamma_+)/(\Gamma_- + \Gamma_+)$ is related to the angle $\gamma$ through the relationship

$$A = \frac{2 r_B r_D \sin(\delta_B + \delta_D) \sin \gamma}{r_B^2 + r_D^2 + 2 r_B r_D \cos(\delta_B + \delta_D) \cos \gamma}, \tag{2.16}$$

where $r_D$ indicates the ratio of $D$-decay amplitudes,

$$r_D = \left| \frac{A(D^0 \to f)}{A(\overline{D}^0 \to f)} \right|, \tag{2.17}$$

and $r_B$ the ratio of $B$-decay amplitudes,

$$r_B = \left| \frac{A(B^- \to \overline{D}^0 K^-)}{A(B^- \to D^0 K^-)} \right| = \left| \frac{A(B^+ \to D^0 K^+)}{A(B^+ \to \overline{D}^0 K^+)} \right|, \tag{2.18}$$

which is usually a small parameter ($r_B \ll 1$) due to the CKM hierarchy. The strong phase difference between the $B$ amplitudes is denoted by $\delta_B$, the corresponding phase difference between $D$ amplitudes by $\delta_D$. The values and methods to extract the $r_{B,D}$ parameters, and the related phases, depend upon the specific decay channel chosen to determine $\gamma$. All the proposed methods have been pursued in multiple final states. The $\gamma$ measurements from charged $B$ decays currently dominate the world averages, as they involve time independent charge asymmetry measurements and there are several final states accessible.

In order to illustrate the TD method, we will use the most recent study of the decay $B_s^0 \to D_s^{+\pm} h^{\pm} \pi^{\pm} \pi^{\mp}$ decays, where $h$ here indicates a kaon or pion, reported by the LHCb collaboration [111]. A fit to the decay-time spectrum of these decays yields the parameter $\gamma - 2\beta_s$, where $\beta_s$ is the CP violating phase in $B_s^0$ decays discussed in section 2.3.4. Figure 2.5 shows the decay-time distribution and mixing asymmetries together with the fit projections, they obtain $\gamma - 2\beta_s = (42^{+19+6}_{-13-2})°$. The current value of $2\beta_s$ is $(0.48 \pm 0.29)°$, thus its uncertainty does not yet affect the precision of this $\gamma$ determination.

Many combinations of different subsets of measurements have been implemented, with the goal of obtaining the best estimate of the angle $\gamma$. In particular, the LHCb experiment [112], which has currently performed the broadest study of this CKM angle, has reported a combination of $\gamma$ measurements, which includes the TD method from both $B^0$ and $B_s^0$ decays; they obtain $\gamma(\text{LHCb}) = (67 \pm 4)°$ [112]. They use a frequentist treatment, starting from a likelihood function built from the product of probability

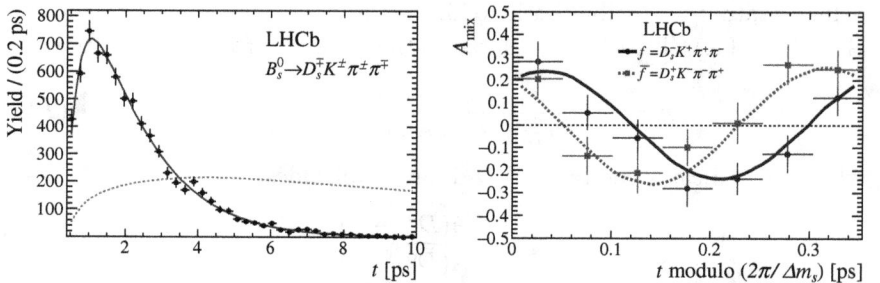

Fig. 2.5: Decay-time distribution [111] of (left) background-subtracted $B_s^0 \to D_s^{\pm} h^{\pm} \pi^{\pm} \pi^{\mp}$ candidates and (right) dilution-weighted mixing asymmetry along with the model-independent fit projections (lines). The decay-time acceptance (left) is overlaid in an arbitrary scale (dashed line).

density functions (PDFs) $f_i$, of the experimental observables $A_i$, where $\vec{A}_i^{\text{obs}}$ are the measured values from an input analysis $i$, and $\vec{\alpha}$ is the set of parameters. For each of the inputs it is assumed that the observables follow a Gaussian distribution

$$f_i(\vec{A}_i^{\text{obs}}|\vec{\alpha}) \propto \exp\left(-\frac{1}{2}(\vec{A}_i(\vec{\alpha}) - \vec{A}_i^{\text{obs}})^T V_i^{-1} (\vec{A}_i(\vec{\alpha}) - \vec{A}_i^{\text{obs}})\right), \quad (2.19)$$

where $V_i$ is the experimental covariance matrix, which includes statistical and systematic uncertainties and their correlations. Note that systematic correlations between statistically independent measurements are assumed to be zero. A $\chi^2$ function is defined as $-2ln\mathcal{L}(\vec{\alpha})$. The best-fit point is given by the global minimum of the $\chi^2$ function. To evaluate the confidence level CL for a given value of the parameter (e.g. $\gamma = \gamma_i$), the value of the $\chi^2$ function at the new minimum is considered ($\chi^2(\vec{\alpha}'(\gamma_i))$). Then $\Delta\chi^2 \equiv \chi^2_{\text{min}} - \chi^2(\vec{\alpha}'(\gamma_i))$ is defined, and a p-value, or 1-CL, is evaluated with a simulation procedure described in Ref. [113]. The full LHCb combination involves 124 observables and 44 free parameters, and the value of the $\chi^2$ at the global minimum is 74.3, this corresponds to a p-value of 68.7%. They obtain $\gamma = (67\pm4)°$, where the quoted uncertainty corresponds to 68.3% CL. The corresponding averages for individual $B$ species are $\gamma(B^+) = (64^{+4}_{-5})°$, $\gamma(B^0) = (82^{+8}_{-9})°$, and $\gamma(B^0_s) = (82^{+17}_{-20})°$.

Information from Belle, BaBar, CDF and LHCb also has been combined into a world average [74] determined with a frequentist method similar to the one described above. They obtain $\gamma(\text{PDG2021}) = (66.2^{+3.4}_{-3.6})°$. Figure 2.6 shows the results of the $\gamma$ combinations of all the experiments

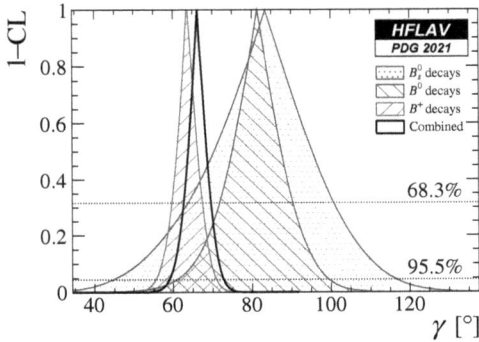

Fig. 2.6: One dimensional 1-CL scan showing the results of the $\gamma$ combination of all the experimental data currently available. The components corresponding to the various $B$-meson species are also reported.

currently available, where the 1-CL scans are performed aggregating data depending upon the species of the decaying $B$ meson. Global fits, using information from other flavor observables, have been performed with two different approaches that will be described in section 2.4.1. They use other data to predict $\gamma$ to be $(66.4 \pm 2.0)°$ from the UT fit group [114] or $(67.7^{+1.0}_{-2.7})°$ from the CKM fitter group [115]. Thus, currently, the measured value of $\gamma$ is consistent with expectations. Note that the world average of direct determinations of $\gamma$ has moved somewhat from the 2019 combination, which found $\gamma = (71.1^{+4.6}_{-5.3})°$ [74].

## 2.3 $B$–$\bar{B}$ mixing and CP violation

The non-vanishing amplitude mixing the quasi-stable neutral pseudoscalar mesons ($M \equiv^0 B_s$, $B_d^0$, $D^0$, or $K^0$) with the corresponding anti-mesons induces time-dependent oscillations between these states. An initially produced $M^0$ or $\bar{M}^0$ evolves in time into a superposition of $M^0$ and $\bar{M}^0$ states.

### 2.3.1 Time evolution of neutral B mesons

We will now focus on the time evolution of neutral $B$ mesons. Denoting by $|B^0(t)\rangle$ (or $|\bar{B}^0(t)\rangle$) the state vector of a $B$ meson which is tagged as a $B^0$ (or $\bar{B}^0$) at time $t = 0$, the time evolution of these states is governed by the following equation:

$$i\frac{d}{dt}\begin{pmatrix} |B^0(t)\rangle \\ |\bar{B}^0(t)\rangle \end{pmatrix} = \left(M - i\frac{\Gamma}{2}\right)\begin{pmatrix} |B^0(t)\rangle \\ |\bar{B}^0(t)\rangle \end{pmatrix} \equiv \Sigma \begin{pmatrix} |B^0(t)\rangle \\ |\bar{B}^0(t)\rangle \end{pmatrix}, \qquad (2.20)$$

where the mass-matrix $M$ and the decay matrix $\Gamma$ are $t$-independent Hermitian $2 \times 2$ matrices. CPT invariance implies that $M_{11} = M_{22}$ and $\Gamma_{11} = \Gamma_{22}$. Even the smallest off-diagonal element lifts the degeneracy between flavor eigenstates and gives rise to flavor oscillations. The off-diagonal elements

$$M_{12} = M_{21}^* \equiv \frac{\Sigma_{12} + \Sigma_{12}^*}{2} \quad \text{and} \quad \frac{\Gamma_{12}}{2} \equiv \frac{\Sigma_{12} - \Sigma_{12}^*}{2} \qquad (2.21)$$

are the dispersive ($M_{12}$) and absorptive ($\Gamma_{12}$), or on-shell, components of the box diagram shown in Fig. 1.2. The former is highly sensitive to NP, whereas the latter is expected to be less affected by possible non-standard dynamics.

The mass eigenstates are the eigenvectors of $M - i\Gamma/2$. We express them in terms of the flavor eigenstates as

$$|B_L\rangle = p|B^0\rangle + q|\bar{B}^0\rangle, \quad |B_H\rangle = p|B^0\rangle - q|\bar{B}^0\rangle, \tag{2.22}$$

with $|p|^2 + |q|^2 = 1$. Note that, in general, $|B_L\rangle$ and $|B_H\rangle$ are not orthogonal to each other. The time evolution of the mass eigenstates is governed by the two eigenvalues $M_H - i\Gamma_H/2$ and $M_L - i\Gamma_L/2$:

$$|B_{H,L}(t)\rangle = e^{-(iM_{H,L}+\Gamma_{H,L}/2)t}|B_{H,L}(t=0)\rangle. \tag{2.23}$$

For later convenience it is also useful to define

$$m = \frac{M_H + M_L}{2}, \quad \Gamma = \frac{\Gamma_L + \Gamma_H}{2}, \quad \Delta M = M_H - M_L, \quad \Delta\Gamma = \Gamma_L - \Gamma_H. \tag{2.24}$$

In the absence of CP violation in mixing, the mass eigenstates are also CP eigenstates. If we adopt the convention that the neutral $B$ flavor eigenstates transform under CP as

$$\text{CP}|B^0\rangle = -|\bar{B}^0\rangle, \tag{2.25}$$

the two CP eigenstates can be defined as

$$|B^0{}_{\text{even}}\rangle = \frac{1}{\sqrt{2}}(|B^0\rangle - |\bar{B}^0\rangle), \tag{2.26}$$

$$|B^0{}_{\text{odd}}\rangle = \frac{1}{\sqrt{2}}(|B^0\rangle + |\bar{B}^0\rangle), \tag{2.27}$$

which coincide with $|B_L\rangle$ and $|B_H\rangle$ in the limit $|q/p| = 1$ (corresponding to the absence of CP violation in mixing) [35].

The time evolution of initially tagged $B^0$ or $\bar{B}^0$ states is

$$|B^0(t)\rangle = e^{-imt}\,e^{-\Gamma t/2}\left[f_+(t)\,|B^0\rangle + \frac{q}{p}\,f_-(t)\,|\bar{B}^0\rangle\right],$$
$$|\bar{B}^0(t)\rangle = e^{-imt}\,e^{-\Gamma t/2}\left[\frac{p}{q}\,f_-(t)\,|B^0\rangle + f_+(t)\,|\bar{B}^0\rangle\right], \tag{2.28}$$

where

$$f_+(t) = \cosh\frac{\Delta\Gamma t}{4}\cos\frac{\Delta M t}{2} - i\sinh\frac{\Delta\Gamma t}{4}\sin\frac{\Delta M t}{2}, \tag{2.29}$$

$$f_-(t) = -\sinh\frac{\Delta\Gamma t}{4}\cos\frac{\Delta M t}{2} + i\cosh\frac{\Delta\Gamma t}{4}\sin\frac{\Delta M t}{2}. \tag{2.30}$$

In both $B_s^0$ and $B_d^0$ systems the following hierarchies holds: $|\Gamma_{12}| \ll |M_{12}|$, hence $\Delta\Gamma \ll \Delta M$. They are experimentally verified and can be traced back to the fact that $|\Gamma_{12}|$ is a genuine long-distance $\mathcal{O}(G_F^2)$ effect, which does not share the large $m_t$ enhancement of $|M_{12}|$ (see section 2.3.5). Taking into account this hierarchy leads to the following approximate expressions for the quantities appearing in the time-evolution formulae in terms of $M_{12}$ and $\Gamma_{12}$:

$$\Delta M = 2\,|M_{12}| \left[1 + \mathcal{O}\left(\left|\frac{\Gamma_{12}}{M_{12}}\right|^2\right)\right], \qquad (2.31)$$

$$\Delta\Gamma = 2\,|\Gamma_{12}| \cos\phi_{12} \left[1 + \mathcal{O}\left(\left|\frac{\Gamma_{12}}{M_{12}}\right|^2\right)\right], \qquad (2.32)$$

$$\frac{q}{p} = -e^{-i\phi_M} \left[1 - \frac{1}{2}\left|\frac{\Gamma_{12}}{M_{12}}\right|\sin\phi_{12} + \mathcal{O}\left(\left|\frac{\Gamma_{12}}{M_{12}}\right|^2\right)\right], \qquad (2.33)$$

where $\phi_M$ is the phase of $M_{12}$. Similarly we can define $\phi_\Gamma$ as the phase of $\Gamma_{12}$. Note that $\phi_B$ and $\phi_\Gamma$ thus defined are unphysical phase-convention quantities, whereas the mixing phase $\phi_{12}$

$$\phi_{12} \equiv \arg\left(-\frac{M_{12}}{\Gamma_{12}}\right) = \pi + \phi_B - \phi_\Gamma, \qquad (2.34)$$

is a physical quantity that can be measured.

Taking into account the above results, the time-dependent decay rates of an initially tagged $B^0$ into some final state $f$ can be written as [116]

$$\Gamma(B^0 \to f) = \mathcal{N}_0 |\mathcal{A}_f|^2 (1 + |\lambda_f|^2)(1 + a_{\text{FS}}) e^{-\Gamma t}$$
$$\times \left\{ \frac{\cosh(\Delta\Gamma t/2)}{2} + \frac{1 - |\lambda_f|^2}{1 + |\lambda_f|^2} \times \frac{\cos(\Delta M t)}{2} \right.$$
$$\left. - \frac{\text{Re}(\lambda_f)}{1 + |\lambda_f|^2} \sinh(\Delta\Gamma t/2) - \frac{\text{Im}(\lambda_f)}{1 + |\lambda_f|^2} \sin(\Delta M t) \right\}, \quad (2.35)$$

where $\mathcal{A}_f$ represents the decay amplitude from the flavor eigenstate to the final state $f$,

$$\mathcal{A}_f = \langle f|\mathcal{L}_{\Delta F=1}|B^0\rangle, \qquad \bar{\mathcal{A}}_f = \langle f|\mathcal{L}_{\Delta F=1}|\bar{B}^0\rangle, \qquad (2.36)$$

and $\lambda_f$, whose imaginary part is related to CP violation in the interference between mixing and decay, is defined as

$$\lambda_f = \frac{q}{p} \frac{\bar{A}_f}{A_f} \approx -e^{-i\phi_M} \frac{\bar{A}_f}{A_f} \left[ 1 - \frac{1}{2} \left| \frac{\Gamma_{12}}{M_{12}} \right| \sin\phi \right]. \tag{2.37}$$

The factor $\mathcal{N}_0$ is the flux normalization and $a_{\mathrm{FS}}$, related to CP violation in mixing, is defined as

$$a_{\mathrm{FS}} = -2 \left( \left| \frac{q}{p} \right| - 1 \right). \tag{2.38}$$

For completeness, we quote the time evolution of a meson $B^0$ or $\bar{B}^0$ to the CP conjugate of the state $f$, namely $\bar{f}$:

$$\Gamma\left[ B^0(t) \to \bar{f} \right] = N_f \left| \bar{A}_{\bar{f}} \right|^2 \left( 1 + |\lambda_{\bar{f}}|^{-2} \right) (1 - a_{\mathrm{FS}}) e^{-\Gamma t}$$

$$\times \left\{ \frac{\cosh\left(\frac{\Delta\Gamma}{2}t\right)}{2} - \frac{1 - |\lambda_{\bar{f}}|^{-2}}{1 + |\lambda_{\bar{f}}|^{-2}} \frac{\cos(\Delta M t)}{2} \right.$$

$$\left. - \frac{\mathrm{Re}(\lambda_{\bar{f}}^{-1})}{1 + |\lambda_{\bar{f}}|^{-2}} \sinh\left(\frac{\Delta\Gamma}{2}t\right) + \frac{\mathrm{Im}(\lambda_{\bar{f}}^{-1})}{1 + |\lambda_{\bar{f}}|^{-2}} \sin(\Delta M t) \right\}, \tag{2.39}$$

$$\Gamma\left[ \bar{B}^0(t) \to \bar{f} \right] = N_f \left| \bar{A}_{\bar{f}} \right|^2 \left( 1 + |\lambda_{\bar{f}}|^{-2} \right) e^{-\Gamma t}$$

$$\times \left\{ \frac{\cosh\left(\frac{\Delta\Gamma}{2}t\right)}{2} + \frac{1 - |\lambda_{\bar{f}}|^{-2}}{1 + |\lambda_{\bar{f}}|^{-2}} \frac{\cos(\Delta M t)}{2} \right.$$

$$\left. - \frac{\mathrm{Re}(\lambda_{\bar{f}}^{-1})}{1 + |\lambda_{\bar{f}}|^{-2}} \sinh\left(\frac{\Delta\Gamma}{2}t\right) - \frac{\mathrm{Im}(\lambda_{\bar{f}}^{-1})}{1 + |\lambda_{\bar{f}}|^{-2}} \sin(\Delta M t) \right\}. \tag{2.40}$$

The above formulas can be used to extract the observables $\Delta M$, $\Delta\Gamma$, real and imaginary parts of $\lambda_f(\lambda_{\bar{f}})$, and the CP-violating parameter $a_{\mathrm{FS}}$.

### 2.3.2 *Measurements of $\Delta M$*

Flavor oscillations in neutral $B$ mesons have been studied experimentally in the $B^0$ and $B_s^0$ systems [12]. Information about the $B^0$ system comes from $e^+e^-$ b-factories and hadron machines. Figure 2.7 shows the legacy measurement of the parameters $\Delta M_d$ and $\Delta M_s$ from the $e^+e^-$ b-factories

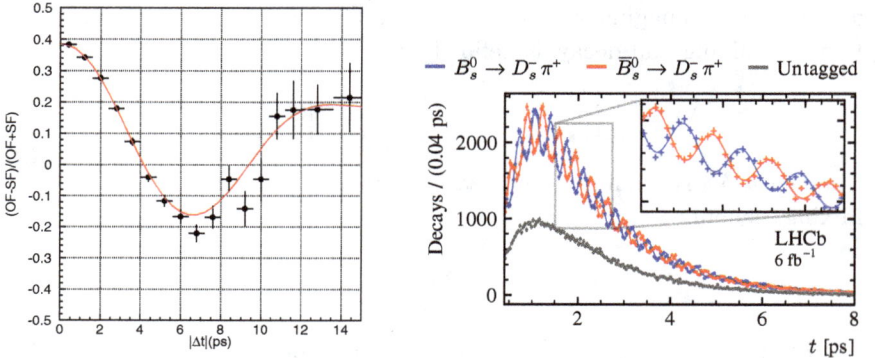

Fig. 2.7: Time-dependent flavor asymmetry (left) for $B^0$-eigenstates [117]. The curve superimposed to the $B^0$ flavor oscillation data is the result of an unbinned maximum-likelihood fit; distribution of the (right) decay time of the $B_s^0 \to D_s^+\pi^-$ signal decays $B_s^0$-eigenstates [118].

and LHCb respectively. The parameter $\Delta M_d$ has been measured by several experiments, including the LEP experiments, D0 and CDF at the Tevatron, the $e^+e^-$ b-factories, and LHCb [74].

The rapid $B_s^0$ oscillations and its heavier mass make the measurement of $\Delta M_s$ more challenging. The first data came from CDF [119], followed by a much more precise determination by the LHCb experiment [120]. Recently, the LHCb collaboration has reported a measurement of $\Delta M_s$ that improved the previous precision by a factor of two. The decay used is $B_s^0 \to D_s^-\pi^+$ and the data are shown in Fig. 2.8. The result is $\Delta M_s = (17.7683 \pm 0.0051 \pm 0.0032)$ ps$^{-1}$. This result is combined with previous LHCb measurements to determine $\Delta M_s = (17.7656 \pm 0.0057)$ ps$^{-1}$, which corresponds to an impressive accuracy of 0.03%.

### 2.3.3  *CP violation in mixing*

CP violation is observed as a difference in rate between two CP conjugate processes. A necessary condition for the manifestation of CP violation is a process mediated by two amplitudes that depend upon two different CKM couplings. In $B^0$ decays, there are three different processes to be considered: CP violation in mixing, CP violation in decays, generally occurring via the interference between two contributions to the decay amplitude, such as tree and penguin diagrams, and CP violation that arises from interference between mixing and decay. We first consided CP violation in mixing, which is described by the weak phase $\phi_{12}$ introduced before. It can be measured

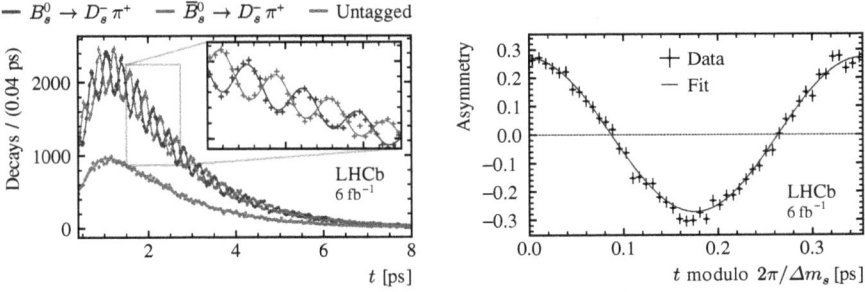

Fig. 2.8: (left) Decay time distribution of the measured $B_s^0 \to D_s^+\pi^-$ signal [121], showing the spectra of events tagged as $B_s^0 \to D_s^-\pi^+$ (blue/dark-grey), $\overline{B}_s^0 \to D_s^-\pi^+$ (red/light grey) and untagged events (lower curve); (right) decay time asymmetry between mixed and unmixed signal decays.

directly via CP asymmetries of flavor-specific decays, defined as

$$a_{\text{FS}} = \frac{\Gamma(\overline{B}^0 \to f) - \Gamma(B^0 \to \bar{f})}{\Gamma(\overline{B}^0 \to f) + \Gamma(B^0 \to \bar{f})}. \tag{2.41}$$

Indeed, inserting the time evolution of the $B^0$ meson, leads to

$$a_{\text{FS}} = -2\left(\left|\frac{q}{p}\right| - 1\right) = \text{Im}\left(\frac{\Gamma_{12}}{M_{12}}\right) = \left|\frac{\Gamma_{12}}{M_{12}}\right| \sin\phi_{12}. \tag{2.42}$$

Typically the flavor of the initial $B$ hadron is not observed, to avoid the penalty incurred by the tagging efficiency, and $a_{\text{SL}}$ is extracted from the time-independent untagged asymmetry

$$a_{\text{FS,unt}} \equiv \frac{\Gamma[f,t] - \Gamma[\bar{f},t]}{\Gamma[f,t] + \Gamma[\bar{f},t]} = \frac{a_{\text{FS}}}{2} - \frac{a_{\text{FS}}}{2}\frac{\cos\Delta M t}{\cosh\Delta\Gamma t/2}, \tag{2.43}$$

where

$$\Gamma[f,t] \equiv \Gamma(B^0 \to f) + \Gamma(\overline{B}^0 \to f). \tag{2.44}$$

The above equations are only valid if we start out with equal production of $B_s^0$, labeled $N$, and $\overline{B}_s^0$, labeled $\overline{N}$. If we allow for $N \neq \overline{N}$, we have

$$\Gamma(M(t) \to f) = N|A_f|^2 e^{-\Gamma t}\left\{\frac{1}{2}\cosh\frac{\Delta\Gamma t}{2} + \frac{1}{2}\cos(\Delta m\,t)\right\}$$

$$\tag{2.45}$$

$$\Gamma(\overline{M}(t) \to \bar{f}) = \overline{N}|A_f|^2\frac{1}{1-a_{sl}}e^{-\Gamma t}\left\{\frac{1}{2}\cosh\frac{\Delta\Gamma t}{2} - \frac{1}{2}\cos(\Delta m\,t)\right\}.$$

Summing we have

$$\Gamma[f,t] \equiv \Gamma(M(t) \to f) + \Gamma(\overline{M}(t) \to f)$$
$$= \frac{|A_f|^2}{2} \left\{ \left( N + \frac{\overline{N}}{1-a_{sl}} \right) \cosh\frac{\Delta\Gamma t}{2} + \left( N - \frac{\overline{N}}{1-a_{sl}} \right) \cos(\Delta m\, t) \right\}.$$

$$(2.46)$$

Similarly

$$\Gamma(M(t) \to \overline{f}) = N\,|A_f|^2\,(1-a_{sl})e^{-\Gamma t} \left\{ \frac{1}{2}\cosh\frac{\Delta\Gamma t}{2} + \frac{1}{2}\cos(\Delta m\, t) \right\}$$

$$\Gamma(\overline{M}(t) \to \overline{f}) = \overline{N}\,|A_f|^2\,e^{-\Gamma t} \left\{ \frac{1}{2}\cosh\frac{\Delta\Gamma t}{2} - \frac{1}{2}\cos(\Delta m\, t) \right\} \qquad (2.47)$$

and the sum is given by

$$\Gamma[\overline{f},t] \equiv \Gamma(M(t) \to \overline{f}) + \Gamma(M(t) \to \overline{f})$$
$$= \frac{|A_f|^2}{2} \left\{ \left( N(1-a_{\mathrm{FS}}) + \overline{N} \right)\cosh\frac{\Delta\Gamma t}{2} + \left( N(1-a_{sl}) - \overline{N} \right)\cos(\Delta m\, t) \right\}.$$

$$(2.48)$$

The production asymmetry, $a_p$, is defined as

$$a_p = \frac{N - \overline{N}}{N + \overline{N}}. \qquad (2.49)$$

To first order in $a_{\mathrm{FS}}$ we have

$$\frac{\Gamma[f,t] - \Gamma[\overline{f},t]}{\Gamma[f,t] + \Gamma[\overline{f},t]} = \frac{a_{\mathrm{FS}}}{2} + \left[ a_p - \frac{a_{\mathrm{FS}}}{2} \right] \frac{e^{-\Gamma t}\cos(\Delta m\, t)\epsilon(t)}{e^{-\Gamma t}\cosh\frac{\Delta\Gamma t}{2}\epsilon(t)}, \qquad (2.50)$$

where $\epsilon(t)$ is the time-dependent acceptance function, a smooth function starting at zero for zero $B_s^0$ decay time, and determined from Monte Carlo simulation.

For a time-independent measurement we integrate the equations involving $N$ and $\overline{N}$ over time finding

$$\frac{\Gamma[f] - \Gamma[\overline{f}]}{\Gamma[f] + \Gamma[\overline{f}]} = \frac{a_{\mathrm{FS}}}{2} + \left[ a_p - \frac{a_{\mathrm{FS}}}{2} \right] \frac{\int_{t=0}^{\infty} e^{-\Gamma t}\cos(\Delta m\, t)\epsilon(t)dt}{\int_{t=0}^{\infty} e^{-\Gamma t}\cosh\frac{\Delta\Gamma t}{2}\epsilon(t)dt}. \qquad (2.51)$$

The integral in the numerator of the last term is vanishingly small for $B_s^0$ decays due to the rapid $B_s^0$ oscillations. For $B^0$ decays it must be carefully evaluated. The denominator is of $\mathcal{O}(1)$, and $a_p$ is of $\mathcal{O}(1\%)$, making its effect negligible in $B_s^0$ decays.

The flavor-specific asymmetry $a_{\mathrm{FS}}$ is expected to be highly suppressed in the SM, and thus its measurement represents an effective null-test of the SM. While in principle any flavor-specific final state can be used to measure $a_{\mathrm{FS}}$, in practice, the goal of minimizing statistical uncertainties oriented experimental searches towards the use of semileptonic decays. The quantity measured is thus generally referred to as $a_{\mathrm{SL}}$. The SM predictions for $B_d$ and $B_s$ mesons are [116]:

$$a_{\mathrm{SL}}^{d,\mathrm{SM}} = -(4.73 \pm 0.42) \times 10^{-4},$$
$$a_{\mathrm{SL}}^{s,\mathrm{SM}} = +(2.22 \pm 0.17) \times 10^{-5}. \tag{2.52}$$

Although the measurement is conceptually simple, the fact that the expected asymmetry is tiny demands that extreme care be taken in assessing any potential source of asymmetry such as production dynamics, background sources, and detection asymmetry. For example, the measured corrected yields for the final states shown in Eq. (2.43) allow us to construct a ratio often defined $A_{raw}$. In order to determine $a_{\mathrm{SL}}$, it is necessary to measure the detection asymmetry $A_d$, accounting for difference in efficiencies between charged particles of opposite sign. Lastly, there may be some production asymmetry to be accounted for [116]. While the smallness of expected SM value poses experimental challenges, it also implies that this is a very good place to look for New Physics, if the systematic uncertainties can be kept small. It will be difficult to have them small enough to see the SM predicted values.

The quantity $a_{\mathrm{SL}}^d$ has been measured, either in decay-time-integrated analyses at CLEO [122, 123], BaBar [124–126], D0 [127], and LHCb [128] or in decay-time-dependent analyses at OPAL [129], ALEPH [130], BaBar [124, 125, 131] and Belle [132]. The quantity $a_{\mathrm{SL}}^s$ has been measured in decay-time-integrated analyses at D0 [133] and LHCb [134]. Table 2.5 summarizes the current status of the world's knowledge,

### 2.3.4 *CP violation in the interference between mixing and decay amplitudes*

An important class of CP-violating effects arise from interference between mixing and decay amplitudes, sometimes it is referred to as *mixing-induced CP violation*. These effects are studied considering final states which are

Table 2.5: Measurements of CP violation in $B^0$ and $B^0_s$ mixing. Some averages incorporate some results converted to an $a^{d,s}_{\text{SL}}$ by the HFLAV averaging group [74].

| | $a^d_{\text{SL}}$ | [× |
|---|---|---|
| B-factory average | $-0.0019 \pm 0.0027$ | |
| D0 | $+0.0068 \pm 0.0045 \pm 0.0014$ | |
| LHCb | $-0.0002 \pm 0.0019 \pm 0.0030$ | |
| Average | $-0.0001 \pm 0.0020$ | |
| | $a^s_{\text{SL}}$ | |
| D0 | $-0.0112 \pm 0.0076$ | |
| LHCb | $+0.0039 \pm 0.0033$ | |

accessible both to the $B$ and $\bar{B}$ hadrons, which allow us to determine the parameter $\lambda_f$ defined in Eq. (2.37).

In general, the decay amplitudes may have multiple contributions from multiple CKM couplings. Their general structure can be decomposed as

$$\mathcal{A}_f = \Sigma_k \mathcal{A}_k e^{i(\phi^{\text{strong}}_k + \phi^{\text{CKM}}_k)}, \qquad (2.53)$$

$$\bar{\mathcal{A}}_f = -\Sigma_k \mathcal{A}_k e^{i(\phi^{\text{strong}}_k - \phi^{\text{CKM}}_k)}. \qquad (2.54)$$

For generic final states, $\lambda_f$ is a quantity that is difficult to evaluate. It becomes particularly simple when $f$ is a CP eigenstate, $\text{CP}|f\rangle = \eta_f|f\rangle$, and a single weak phase dominates the decay amplitude. In such case $\bar{\mathcal{A}}_f/\mathcal{A}_f$ is a pure phase factor ($|\bar{\mathcal{A}}_f/\mathcal{A}_f| = 1$), determined by the difference between the weak phase of the decay amplitude and the phase of the $B$–$\bar{B}$ mixing amplitude:

$$\lambda_f|_{\text{CP-eigen.}} = \eta_f \frac{q}{p} e^{-2i\phi_A}, \quad \mathcal{A}_f = |\mathcal{A}_f| e^{i\phi_A}, \quad \eta_f = \pm 1. \qquad (2.55)$$

In the case of $B_d$ decays, the cleanest example is the $|J/\psi K_S\rangle$ final state. In this case the final state is a CP eigenstate and the decay amplitude is real (to a very good approximation) in the standard CKM phase convention. Indeed the underlying partonic transition is dominated by the Cabibbo-allowed tree-level process $b \to c\bar{c}s$, which has a vanishing phase in the standard CKM phase convention, and also the leading one-loop corrections (top-quark penguins) have the same vanishing weak phase. The decay diagram in shown in Fig. 1.1 where the $c$ and $\bar{c}$ quarks form a $J/\psi$ and the $s$ and $\bar{d}$ quarks form a $K_S$. Since in the $B^0$ system we can safely neglect

$\Gamma_{12}/M_{12}$, this implies

$$\lambda^{B_b}_{J/\psi K_s} = -e^{-i\phi_{B_d}}, \quad \mathrm{Im}\left(\lambda^{B_b}_{J/\psi K_s}\right)_{\mathrm{SM}} = \sin(2\beta). \tag{2.56}$$

Within the the SM, the expression of $\phi_{B_d}$ is nothing but the phase of the CKM combination $(V_{tb}^* V_{td})^2$ that controls the $B$–$\bar{B}$ mixing amplitude [see Eq. (2.63)]. Given the smallness of $\Delta\Gamma_d$, this quantity is easily extracted from the ratio

$$\mathcal{A}^{B \to J/\psi K_s}_{\mathrm{CP}}(t) = \frac{\Gamma[\bar{B}_d(t=0) \to J/\psi K_S(t)] - \Gamma[B^0(t=0) \to J/\psi K_S(t)]}{\Gamma[\bar{B}_d(t=0) \to J/\psi K_S(t)] + \Gamma[B^0(t=0) \to J/\psi K_S(t)]}$$

$$\approx \mathrm{Im}\left(\lambda^{B_b}_{J/\psi K_s}\right)\sin(\Delta M_{B_d} t), \tag{2.57}$$

which can be considered the golden measurement of the CKM angle $\sin 2\beta$.

The analogous golden mode for the $B_s$ meson is the final state $|J/\psi\phi\rangle$. In this case we cannot neglect $\Delta\Gamma_s$, and the final state is not a CP eigenstate. Isolating one of the two CP components of the final state, the CP asymmetry takes the form

$$\mathcal{A}^{B_s \to [J/\psi\phi]\mathrm{CP}}_{\mathrm{CP}}(t) \approx \frac{\sin\phi_s \sin(\Delta M_s t)}{\cos\phi_s \sinh\left(\frac{\Delta\Gamma_s t}{2}\right) - \cosh\left(\frac{\Delta\Gamma_s t}{2}\right)}, \tag{2.58}$$

where, within the SM, $\phi_s$ is given by

$$\phi_s = -\arg\left[\eta_{\mathrm{CP}}\frac{V_{ts}^* V_{tb}}{V_{ts} V_{tb}^*}\frac{V_{cs}^* V_{cb}}{V_{cs} V_{cb}^*}\right]. \tag{2.59}$$

For the $\eta_{\mathrm{CP}} = +1$ component we get

$$\phi_s = -2\beta_s \tag{2.60}$$

where $\beta_s$ is defined as

$$\beta_s = -\arg\left[-\frac{V_{ts}^* V_{tb}}{V_{cs}^* V_{cb}}\right]. \tag{2.61}$$

The above result holds under the hypothesis that subleading penguin topologies in the decay amplitude, with different CKM factors, can be neglected. This holds up to $\pm 1°$ accuracy in Eq. (2.60), below which penguin topologies need to be understood in more detail.

The experimental study of the angle $\phi_s$ has been pursued mainly through the decay $B_s^0 \to J/\psi h^+ h^-$, where $h^+ h^- = K^+ K^-$ or $\pi^+\pi^-$. The LHCb experiment has studied both final states, while most of the other experiments contributing to this measurement are restricting themselves

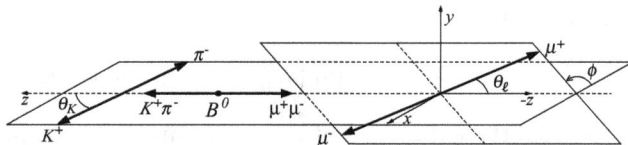

Fig. 2.9: Definition of the helicity angles. For details see text. The plot is taken from [136]. In this figure the angle $\theta_\mu$ is denoted as $\theta_l$.

to the final state $B_s^0 \to J/\psi\phi$. The final state $B_s^0 \to J/\psi K^+ K^-$ can be produced also with $K^+ K^-$ pairs in an S-wave configuration, as pointed out by Stone and Zhang [135]. This S-wave component is CP-odd. In order to extract $\phi_s$, the CP-odd and CP-even components need to be disentangled through an angular analysis of the differential decay width. The relevant angles are shown in Fig. 2.9.

The most recent LHCb result [137] performs a time dependent analysis of $d\Gamma/dt d\Omega$, where $\Omega \equiv (\cos\theta_{K+}\cos\theta_\ell\chi)$. This differential decay width is expressed in terms of 10 components, corresponding to the 4 polarization states $(0, \perp, \|)$ for P-waves, and an S-wave component. Figure 2.10 shows the fit projections of $d\Gamma/dt d\Omega$ for the four variables considered. The most important parameters extracted from this fit are the CP violating phase $\phi_s = -0.0083 \pm 0.041 \pm 0.006$ rad, the mass difference $\Delta\Gamma_s = -0.077 \pm 0.008 \pm 0.003$ ps$^{-1}$ and $\Gamma_s - \Gamma_d = -0.0041 \pm 0.0024 \pm 0.0015$ ps$^{-1}$, where the first uncertainty is statistical and the second is systematic.

Another important decay used by the LHCb experiment to measure $\phi_s$ is $B_s^0 \to J/\psi\pi\pi$. Originally it was considered to be dominated by the final state $B_s^0 \to J/\psi f_0$, with $f_0 \to \pi^+\pi^-$, and thus a predominantly $S$-wave decay. It was later recognized to have a more complex resonant structure [138], and to be compatible with being almost entirely CP-odd. The most recent analysis of this decay includes Run 1 data plus a partial Run 2 data corresponding to an integrated luminosity of 1.9 fb$^{-1}$, they obtain $\phi_s = -0.0057 \pm 0.060 \pm 0.0011$. The two $\phi_s$ measurements presented here are combined with previous LHCb measurements to determine the average $\phi_s = (-0.042 \pm 0.025)$rads.

Measurements of $\phi_s$, $\Delta\Gamma_s$ and $\Gamma_s$ using $B_s^0 \to J/\psi K^+ K^-$ decays, with $J/\psi \to \mu^+\mu^-$, have been previously reported by the D0 [139], CDF [140], ATLAS [141, 142], CMS [143] and LHCb [144] collaborations. The LHCb collaboration has also exploited different decay channels, in addition to $B_s^0 \to J/\psi\pi^+\pi^-$, $B_s^0 \to \psi(2S)\phi$ [145], $B_s^0 \to D_s^+ D_s^-$ [146] and $B_s^0 \to J/\psi K^+ K^-$ for the $K^+ K^-$ invariant-mass region above $1.05$ GeV/$c^2$ [147].

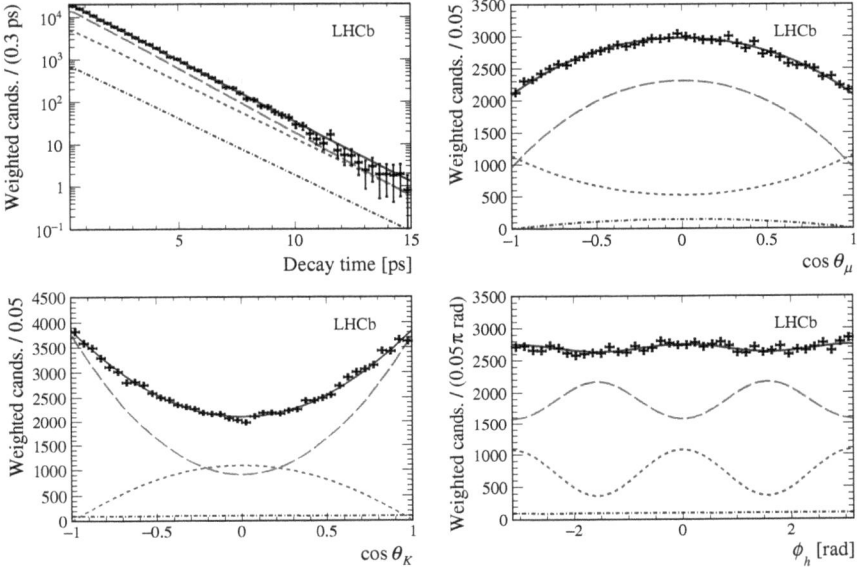

Fig. 2.10: Decay time and helicity angle distributions for background subtracted $B_s^0 \to J/\psi K^+ K^-$ (data points) with the kaon pair consistent with the $\phi$ meson mass. The one-dimensional projections of the PDF at the maximum likelihood point are also shown: total signal contribution (solid lines), CP-even part (long-dashed lines), CP-odd (short-dashed lines), and S-wave (dot-dashed lines).

A new world average including all the results discussed here has been performed by the HFLAV group [74] and is shown in Fig. 2.11.

## 2.3.5  $B^0$ and $B_s^0$ mixing within the SM

Quantities such $M_{ij}$ and $\Gamma_{ij}$, describing $B$ meson oscillations can be calculated in the Standard Model, albeit with the need to tackle strong interaction effects. A first approach is the construction of an effective field theory, based on the hierarchy in the three scales involved in the process of $B$–$\bar{B}$ mixing, $\Lambda_{\rm QCD} \ll m_b \ll m_t \sim m_W$.

In principle, the calculation of $M_{12}$ involves nine different combinations of internal quarks in the box diagram

$$
\begin{aligned}
M_{12} \propto\ & V_{uq}^* V_{ub} F(u,u) + V_{uq}^* V_{cb} F(u,c) + V_{uq}^* V_{tb} F(u,t) + V_{cq}^* V_{ub} F(c,u) \\
& + V_{cq}^* V_{cb} F(c,c) + V_{cq}^* V_{tb} F(c,t) + V_{tq}^* V_{ub} F(t,u) \\
& + V_{tq}^* V_{cb} F(t,c) + V_{tq}^* V_{tb} F(t,t),
\end{aligned}
\tag{2.62}
$$

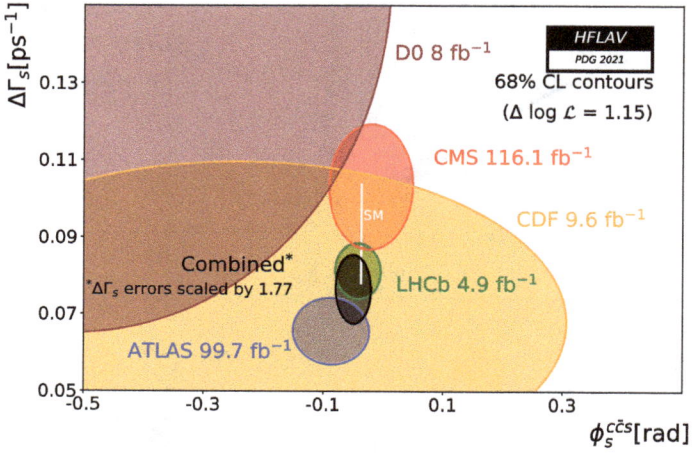

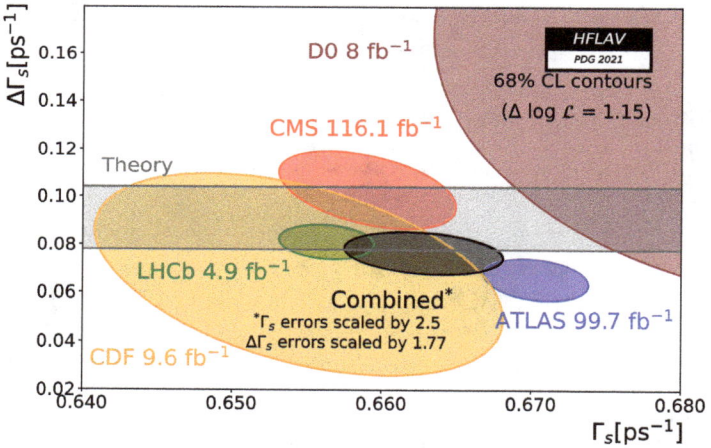

Fig. 2.11: PDG 2021 averages for $\phi_s$, $\Delta\Gamma_s$, and $\Gamma_s$.

where $q = d, s$ depending upon the flavor of the decaying neutral $B$. The functions $F(x, y)$ depend upon the masses of the internal quarks $x$ and $y$, normalized with respect to the $W$ boson mass. Due to the the unitarity of the CKM matrix, contributions to $F(x, y)$ which are quark-mass independent cancel and the amplitude is proportional to quark mass differences. This cancellation holds for all $|\Delta F| = 2$ amplitudes, and works also in the limit of two generations. It is essential to explain the strong suppression of kaon mixing and is known as the "GIM" mechanism, and is the reason that led Glashow, Iliopulos, and Maiani to postulate the existence of charm back in 1970 [148], before its direct discovery in 1974.

Coming back to $B$–$\bar{B}$ mixing, the GIM mechanism implies that the top quark dominates and we can write

$$M_{12} = \frac{G_F^2}{12\pi^3} V_{tq}^* V_{tb} M_W^2 F\left(\frac{m_t^2}{M_{W^2}}\right) f_{B_q}^2 \hat{B}_{B_q} M_{B_q} \eta_B. \qquad (2.63)$$

Here $G_F$ is the Fermi constant, $f_{B_q}$ the decay constant, $M_{B_q}$ the mass of the neutral $B$ meson being considered, and $\hat{B}_{B_q}$ is the so-called *bag parameter*, which allows us to parameterize the $|\Delta B| = 2$ hadronic matrix element as[1]

$$\eta_B(\mu)\langle \bar{B}_q|(\bar{b}_L\gamma_\mu q_L)^2|B_q\rangle = \frac{2}{3}f_{B_q}^2 m_{B_q}^2 \eta_B(\mu) B_q(\mu) = \frac{2}{3}f_{B_q}^2 m_{B_q}^2 \hat{\eta}_B \hat{B}_{B_q}.$$
$$(2.64)$$

The parameter $\hat{\eta}_B$ encodes perturbative QCD corrections, and finally $F$ represents the Inami-Lim function [149] defined as

$$F\left(\frac{m_t^2}{M_{W^2}}\right) \equiv F(z) = \frac{1}{4} + \frac{9}{4(1-z)} - \frac{3}{2(1-z)^2} - \frac{3}{2}\frac{z^2 \ln(z)}{(1-z)^3}. \qquad (2.65)$$

As shown in Eq. (2.64), the product $f_{B_q}^2 \hat{B}_{B_q}$ encodes non-perturbative dynamics. Figure 2.12 summarizes the current lattice evaluation of $\sqrt{\hat{B}_q} f_{B_q}$ and the average performed by the FLAG lattice averaging group [68].

The calculation of the decay rate difference $\Delta\Gamma_{12}$ is more involved. In this case it is necessary to account for the contributions of on-shell particles; thus both the charm and up quarks contribute, but not the top quark. In order to make predictions, one has to rely on HQE [116]: in this approach the total decay rate of a heavy hadron can be expanded in inverse powers of the heavy quark mass as

$$\frac{1}{\tau} = \Gamma = \Gamma_0 + \frac{\Lambda^2}{m_b^2}\Gamma_2 + \frac{\Lambda^3}{m_b^3}\Gamma_3 + \frac{\Lambda^4}{m_b^4}\Gamma_4 + \cdots. \qquad (2.66)$$

The hadronic scale $\Lambda$ is of order $\Lambda_{\text{QCD}}$, but its actual value needs to be determined with a non-perturbative calculations. For hadron lifetimes it turns out that the dominant correction to $\Gamma_0$ is the third term $\Gamma_3$. Each of

---

[1]The $B$ flavor changes from $-1$ to $+1$, or vice-versa.

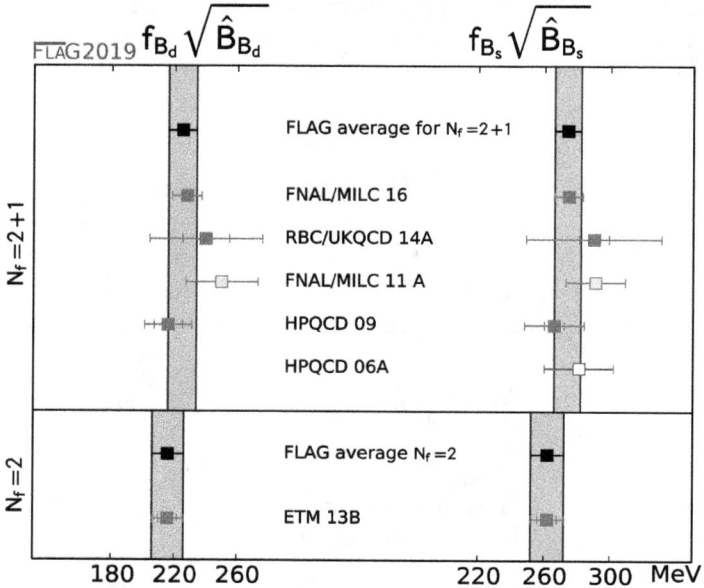

Fig. 2.12: Product of the neutral $B^0$ bag parameters and square-root of the decay constants [68]. $N_f = 2$ refers to lattice calculations using only the $u$ and $d$ quarks, while $N_f = 2 + 1$ also includes the $s$ quark.

the $\Gamma_i$'s can be split up in a perturbative part and non-perturbative matrix elements — it can be formally written as

$$\Gamma_i = \left[\Gamma_i^{(0)} + \frac{\alpha_S}{4\pi}\Gamma_i^{(1)} + \frac{\alpha_S^2}{(4\pi)^2}\Gamma_i^{(2)} + \cdots,\right]\langle O^{d=i+3}\rangle \qquad (2.67)$$

where $\Gamma_i^{(0)}$ denotes the perturbative LO-contribution, $\Gamma_i^{(1)}$ the NLO one and so on; $\langle O^{d=i+3}\rangle$ is the non-perturbative matrix element of $\Delta B = 0$ operators of dimension $i+3$. The mixing quantity $\Gamma_{12}^q$ obeys a similar HQE expansion, but now the operators change the $b$-quantum number by two units, $\Delta B = 2$:

$$\Gamma_{12} = \frac{\Lambda^3}{m_b^3}\Gamma_3 + \frac{\Lambda^4}{m_b^4}\Gamma_4 + \cdots \qquad (2.68)$$

and $\Delta\Gamma_{12}$ can be cast as a power series in the inverse of the heavy $b$ quark mass $m_b$ and the strong coupling constant:

$$\Gamma_{12} = \frac{\Lambda^3}{m_b^3}\left(\Gamma_3^{(0)} + \frac{\alpha_s}{4\pi}\Gamma_3^{(1)} + \cdots\right) + \frac{\Lambda^4}{m_b^4}\left(\Gamma_4^{(0)} + \cdots\right) + \cdots. \qquad (2.69)$$

Table 2.6: SM predictions and experimental data for neutral $B$ mixing observables.

| Observable | World average | SM prediction |
|---|---|---|
| $\Delta m_d$ | $(0.5065 \pm 0.0019)\text{ps}^{-1}$ [74] | $(0.543 \pm 0.029)\text{ ps}^{-1}$ [151] |
| $\Delta \Gamma_d$ | $(0.7 \pm 5.6) \times 10^{-3}$ [74] | $(2.6 \pm 0.4) \times 10^{-3}$ [151] |
| $a_{SL}^d$ | $(-21 \pm 17) \times 10^{-4}$ [74] | $(-4.73 \pm 0.42) \times 10^{-4}$ [151] |
| $\Delta m_s$ | $(17.7656 \pm 0.0057)\text{ ps}^{-1}$ [121] | $(18.66 \pm 0.86)\text{ ps}^{-1}$ [151] |
| $\Delta \Gamma_s$ | $(82.1 \pm 0.5) \times 10^{-2}\text{ ps}^{-1}$ [74] | $(9.1 \pm 1.3) \times 10^{-2}\text{ ps}^{-1}$ [151] |
| $a_{SL}^s$ | $(0.39 \pm 0.26 \pm 0.20)\%$ [134] | $(2.06 \pm 0.18) \times 10^{-5}$ [151] |

where $\Lambda$ represents a hadronic scale assumed to be of the order of $\Lambda_{\text{QCD}}$, but whose actual value needs to be determined with a non-perturbative calculation. Significant theory effort has been spent in evaluation non-leading order corrections. For example, the coefficients of the $\Gamma_{12}^s$ expansion have been evaluated up to $1/m_b^3$ [150].

### 2.3.6 *Summary of neutral B mixing and CP violation observables*

Table 2.6 shows the current world average and quantifies the impact of the b-factories and LHCb.

## 2.4 The global CKM fit and $B$–$\bar{B}$ mixing beyond the SM

### 2.4.1 *Fits to experimental CKM constraints*

The measurements discussed in this chapter can be used to provide important tests of the SM model. There are different approaches to the combination of the observables included in our discussion, sometimes augmented by complementary constraints coming from $K$ decays. In some cases all the available information is included in global fits, to examine whether a single "standard unitarity triangle" is consistent with all the observations so far. Alternatively, only measurements mediated by tree level process or measurements associated with loop diagrams are considered, to discern whether the latter are inconsistent with the former, and perhaps hinting at some tensions with SM expectations. Lastly, specific observables may be selected on the basis of more reliable theoretical predictions.

The challenge in these global fits is a proper treatment of systematic and theoretical uncertainties. Two main approaches form the basis of our current knowledge. The UT fit collaboration [114] uses a Bayesian approach.

Each measured quantity $c_j$ constrains the CKM-triangle parameters $\bar{\rho}$ and $\bar{\eta}$ via a set of ancillary parameters $\vec{x} \equiv [x_1, \ldots, x_N]$, which represent all the experimentally determined or theoretically calculated quantities from which the various $c_j$ depend

$$c_j = c_j(\bar{\rho}, \bar{\eta}; \vec{x}) \tag{2.70}$$

The parameters $\vec{x}$ are affected by experimental and theoretical uncertainties, which, in turn, in both cases can encompass a statistical and systematic component. Bayesian inference is used to determine domains in $\bar{\rho} - \bar{\eta}$ space with a given confidence level. If a PDF describing the uncertainty of a constraint is available, it is used in the fit; for variables for which a range is given, a flat distribution is assumed. The CKM fitter collaboration [115] uses a standard $\chi^2$-like frequentist approach, in addition to the RFit scheme to treat theoretical uncertainties. While there is a vast literature on the relative merit of the two different approaches, most of the fits performed with the same inputs have very consistent results.

Generally, fits referred as "global fits" include all the relevant information available. In some cases, a subset of the measured variables may be chosen on the basis of their physics nature; for example one might consider only variables arising from tree-level processes, or select variables that have more robust theoretical uncertainty evaluations. In the case of the $|V_{ub}|/|V_{cb}|$ constraint, an average value of these parameters is commonly used, which may not be optimal as the tension between inclusive and exclusive determination may have a physics origin or may reflect non-quantified discrepancies. Figure 2.13 shows the results of the unitarity triangle fit using all constraints available in the Summer of 2021 [114]. This analysis includes all the results from $B$ decays, including the magnitude of the quark couplings $V_{ub}$ and $V_{cb}$, the constraints coming from $B^0$ and $B^0_s$ mixing, and the constraint coming from CP violation in $K^0$ decay. The UT fitter group finds $\bar{\rho} = 0.157 \pm 0.012$ and $\bar{\eta} = 0.350 \pm 0.010$ at 95% CL. All the measurements used are consistent with this apex and this global fit can be seen as a triumph of the SM. The CKM fitter group [115], in a fit completed at the end of 2019, obtains $\bar{\rho} = 0.157^{+0.027}_{-0.012}$ and $\bar{\eta} = 0.350^{+0.018}_{-0.016}$ at 95% CL. This suggests that the global CKM fits provide a remarkable triumph of the SM. However, as the experimental inputs become more precise, and the theoretical calculations acquire higher precision and better understanding of their systematic uncertainties, a careful analysis

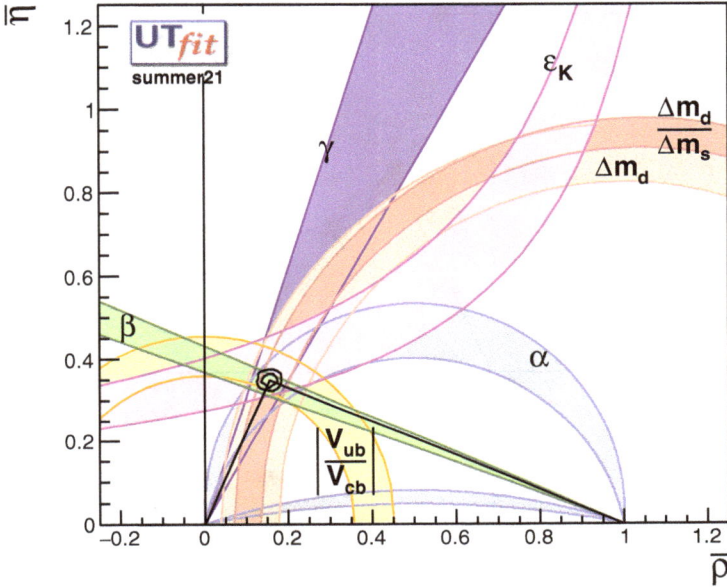

Fig. 2.13: Global SM CKM fits by the UTfit collaboration [114] (Summer 2021).

of fits performed using specific inputs may elucidate subtle effects not yet uncovered in the global fits.

Further insight can be obtained by considering subset of experimental inputs. In particular we can consider only observables associated with $B$ decays, and distinguish the measurements related to tree-level processes and loop-processes. Figure 2.14 shows the two regions for the apex of unitarity triangle determined using only either tree level quantities or loop quantities. The two regions are consistent with the global fits, but small tensions emerge. A further subtlety to be noted is that, in choosing the experimental values of $|V_{ub}|$ and $|V_{cb}|$, a choice needs to be made between which value to use, namely the one derived from exclusive determinations, inclusive determinations, or an average of the two. Figure 2.15 illustrates the unresolved tension in the $|V_{ub}|$ case: the prediction derived by all the other variables included in the global fit is compared with the two direct determinations. It can be seen that the inclusive determination shows a $\sim 2\sigma$ discrepancy with the predicted value. This discrepancy may reflect unquantified systematic uncertainties or may have a physics explanation. In absence of a resolution of this multi-decade old tension a preferred approach

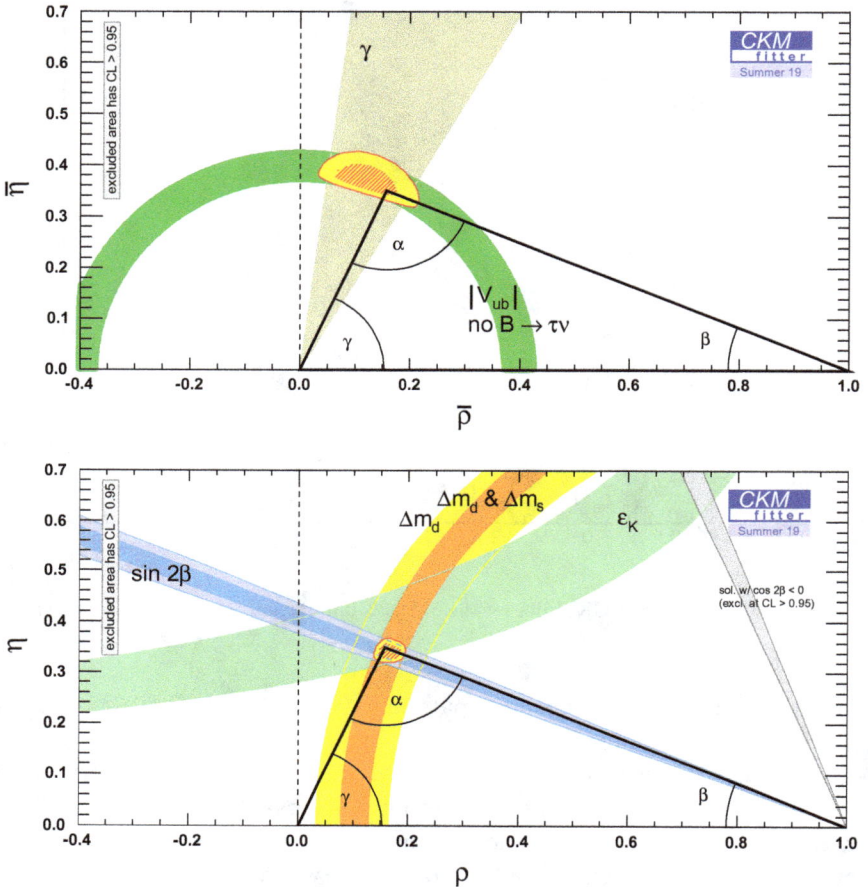

Fig. 2.14: The CKM fit using tree-level observables only (top) and loop-observables only (bottom).

would be to both of these determinations separately, or to average the two results and inflate the uncertainty.

### 2.4.2 *NP bounds from $B_{s,d}$ mixing and the flavor problem*

A very plausible hypothesis is to assume that NP effects are negligible in processes which are dominated by tree-level amplitudes. More precisely, we can assume NP contributions are negligible in tree-level $B$ decays involving only quarks and leptons of the first two generations in the final state. Following this assumption, the determination of the CKM elements $|V_{us}|$,

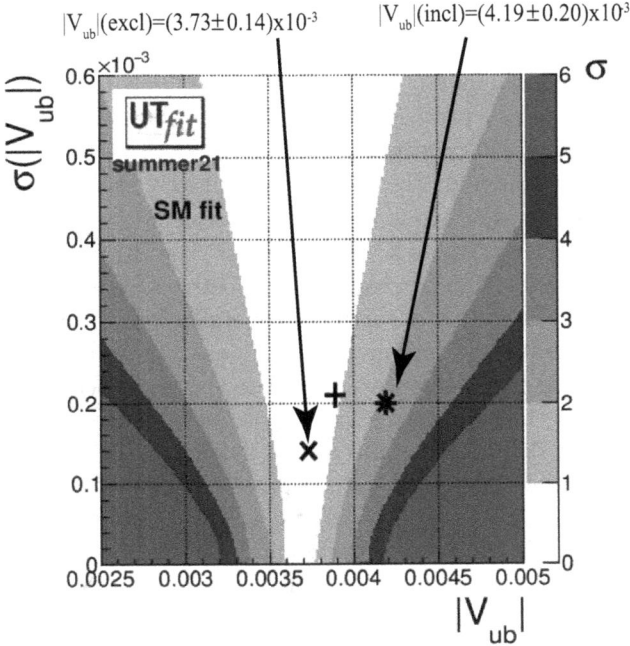

Fig. 2.15: Comparison between the expectations from global UTfit predictions, exclusive $|V_{ub}| = (3.73 \pm 0.14) \times 10^{-3}$, and inclusive $|V_{ub}| = (4.19 \times 0.20) \times 10^{-3}$. The corresponding average is $|V_{ub}| = (3.89 \pm 0.21) \times 10^{-3}$. The UTfit prediction from other observables is $|V_{ub}| = (3.68 \pm 0.10) \times 10^{-3}$.

$|V_{cb}|$, and $|V_{ub}|$, as well as the constraints on $\alpha$ and $\gamma$ are essentially NP free. The value of $\rho$ and $\eta$ obtained under this hypothesis are those reported in Fig. 2.14. We can use these values to predict $B_{s,d}$ mixing within the SM and compare the predictions with data, setting bounds on possible NP contributions. Before doing so, it is worth giving a closer look at the structure of $B_{s,d}$ mixing within the SM.

## $B_{s,d}$ mixing in the the gauge-less limit

We already discussed the theoretical expressions of the mixing amplitudes within the SM. The key feature is that the leading contribution is obtained by the top-quark running inside the loop: the amplitudes grow with $m_t$ and they diverge in the limit $m_t/m_W \to \infty$. This feature is present not only in $B_{s,d}$ mixing: it is present in all down-type $|\Delta F| = 2$ amplitudes (i.e. $B_{s,d}$ mixing and neutral kaon mixing), as well as in the so-called $Z$-penguin amplitudes contributing to rare FCNC transitions of the type $b \to s\ell^+\ell^-$, which we will discuss in the next chapter.

This behavior is apparently strange: it contradicts the expectation that contributions of heavy particles at low energy decouple in the limit where their masses increase. The origin of this effect can be understood by noting that the leading contributions to both amplitudes are generated only by the Yukawa interaction. These contributions can be better isolated in the *gaugeless* limit of the SM, i.e. if we send to zero the gauge couplings. In this limit $m_W \to 0$ and the derivation of the effective Lagrangian discussed in section 1.4.2, for processes occurring at energies well below $m_W$ does not make sense. However, the leading contributions to the effective Lagrangians for $|\Delta F| = 2$ and rare decays are unaffected. This is because the leading contributions to these processes are generated by Yukawa interactions of the type in Fig. 2.16, where the scalar fields are the Goldstone-bosons components of the Higgs field (which are not eaten up by the $W$ in the limit $g \to 0$).

The case of $B_{s,d}$ mixing is particularly simple. Since the top is still heavy, we can integrate it out, obtaining the following result for $\mathcal{L}_{|\Delta B|=2}$:

$$\mathcal{L}^{SM}_{|\Delta B|=2}\Big|_{g_i \to 0} = \frac{G_F^2 m_t^2}{16\pi^2}(V_{tb}^* V_{tq})^2(\bar{b}_L \gamma_\mu q_L)^2 = \frac{[(Y_U Y_U^\dagger)_{bq}]^2}{128\pi^2 m_t^2}(\bar{b}_L \gamma_\mu q_L)^2.$$

(2.71)

Taking into account that $S_0(x) \to x/4$ for $x \to \infty$, it is easy to verify that this result is equivalent to the one in Eq. (2.63) in the large $m_t$ limit. A similar structure holds for the $|\Delta B| = 1$ amplitude contributing to $b \to s\ell^+\ell^-$.

The last expression in Eq. (2.71), which holds in the limit where we neglect the charm Yukawa coupling, shows that the decoupling of the amplitude with the mass of the top is compensated by four powers of the top Yukawa coupling in the numerator. The divergence for $m_t \to \infty$ can thus be understood as the divergence of one of the fundamental couplings of the

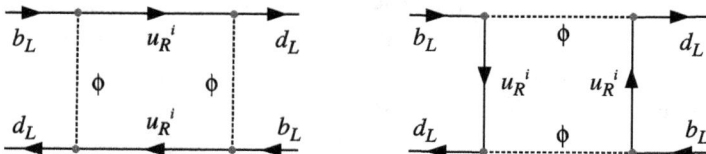

Fig. 2.16: One-loop contributions $|\Delta B| = 2$ amplitudes in the gaugeless limit.

theory. Note also that in the gaugeless limit there is no GIM mechanism: the contributions of the various up-type quarks inside the loops do not cancel each other: they are directly weighted by the corresponding Yukawa couplings, and this is why the top-quark contribution is the dominant one.

This exercise illustrates the key role of the Yukawa coupling in determining the main flavor physics properties within the SM, as advertised in the first chapter. It also illustrates the interplay of flavor and electroweak symmetry breaking in determining the structure of short-distance dominated flavor-changing processes in the SM.

## The flavor problem

Since the magnitude and phase of the mixing amplitudes have been experimentally determined with good accuracy and, within the errors, the results are consistent with the the SM expectations, these measurements imply stringent bounds on NP models.

To translate this information into bounds on the NP scale, let's consider the following generic effective Lagrangian encoding NP effects via a representative set of dimensions-six four-quark operators

$$\Delta\mathcal{L}^{\mathrm{NP}}_{|\Delta F|=2} = \sum_{ab} \frac{c_{ab}}{\Lambda^2} Q^{\mathrm{LL}}_{ab}, \quad Q^{\mathrm{LL}}_{ab} = (\bar{Q}^a_L \gamma^\mu Q^b_L)^2. \tag{2.72}$$

Here $a, b$ are flavor indexes in the basis where the down-type quarks are diagonal (introduced in section 1.1.1). In such basis a given $Q^{\mathrm{LL}}_{ab}$ contribute at the tree-level to a specific down-type meson-antimeson mixing process: $Q^{\mathrm{LL}}_{31}$ contribute at the tree-level to $B_d$ mixing, $Q^{\mathrm{LL}}_{32}$ to $B_s$ mixing, and so on. In the following we keep the discussion generic treating at the same time all down-type meson-antimeson mixing amplitudes.

Since NP is clearly subleading in $B_{s,d}$ mixing and also in the mixing of neutral kaons, the following condition must be satisfied: $|\mathcal{M}^{\mathrm{NP}}_{|\Delta F|=2}| < |\mathcal{M}^{\mathrm{SM}}_{|\Delta F|=2}|$. The latter implies

$$\Lambda < \frac{3.4 \text{ TeV}}{|V^*_{3i} V_{3j}|/|c_{ij}|^{1/2}} < \begin{cases} 9 \times 10^3 \text{ TeV} \times |c_{21}|^{1/2} & \text{from} \quad K^0 - \bar{K}^0 \\ 4 \times 10^2 \text{ TeV} \times |c_{31}|^{1/2} & \text{from} \quad B_d - \bar{B}_d \\ 7 \times 10^1 \text{ TeV} \times |c_{32}|^{1/2} & \text{from} \quad B_s - \bar{B}_s \end{cases} \tag{2.73}$$

A more refined analysis, with complete statistical treatment and separate bounds for the real and the imaginary parts of the various amplitudes, considering also operators with different Dirac structure, and

Table 2.7: Bounds on representative dimension-six $|\Delta F| = 2$ operators, assuming an effective coupling $c_{ab}/\Lambda^2$ (from Ref. [152]). The bounds are quoted on $\Lambda$, setting $|c_{ab}| = 1$, or on $|c_{ab}|$, setting $\Lambda = 1$ TeV. The right column denotes the main observables used to derive these bounds: $\Delta m_i$ indicates mass differences derived from mixing measurements, $\epsilon_K$ CP violation measured in the $K^0$ system, $|q/p|$ and $\phi_D$ CP violation in $D^0$ decays, $S_{\psi K_s}$ CP violation in $B^0 \to J/\psi K^0$, and $S_{\psi\phi}$ CP violation in $B_s^0 \to J/\psi K^+ K^-$ and $J/\psi \pi^+ \pi^-$ decays.

| Operator | Bounds on $\Lambda$ in TeV $|c_{ab} = 1|$ | | Bounds on $|c_{ab}|$ ($\Lambda = 1$ TeV) | | Observables |
|---|---|---|---|---|---|
| | Re | Im | Re | Im | |
| $(\bar{s}_L \gamma^\mu d_L)^2$ | $9.8 \times 10^2$ | $1.6 \times 10^4$ | $9.0 \times 10^{-7}$ | $3.4 \times 10^{-9}$ | $\Delta m_K; \epsilon_K$ |
| $(\bar{s}_R d_L)(\bar{s}_L d_R)$ | $1.8 \times 10^4$ | $3.2 \times 10^5$ | $6.9 \times 10^{-9}$ | $2.6 \times 10^{-11}$ | $\Delta m_K; \epsilon_K$ |
| $(\bar{c}_L \gamma^\mu u_L)^2$ | $1.2 \times 10^3$ | $2.9 \times 10^3$ | $5.6 \times 10^{-7}$ | $1.0 \times 10^{-7}$ | $\Delta m_D; |q/p|, \phi_D$ |
| $(\bar{c}_R u_L)(\bar{c}_L u_R)$ | $6.2 \times 10^3$ | $1.5 \times 10^4$ | $5.7 \times 10^{-8}$ | $1.1 \times 10^{-8}$ | $\Delta m_D; |q/p|, \phi_D$ |
| $(\bar{b}_L \gamma^\mu d_L)^2$ | $6.6 \times 10^2$ | $9.3 \times 10^2$ | $2.3 \times 10^{-6}$ | $1.1 \times 10^{-6}$ | $\Delta m_{B_d}; S_{\psi K_S}$ |
| $(\bar{b}_R d_L)(\bar{b}_L d_R)$ | $2.5 \times 10^3$ | $3.6 \times 10^3$ | $3.9 \times 10^{-7}$ | $1.9 \times 10^{-7}$ | $\Delta m_{B_d}; S_{\psi K_S}$ |
| $(\bar{b}_L \gamma^\mu s_L)^2$ | $1.4 \times 10^2$ | $2.5 \times 10^2$ | $5.0 \times 10^{-5}$ | $1.7 \times 10^{-5}$ | $\Delta m_{B_s}; S_{\psi\phi}$ |
| $(\bar{b}_R s_L)(\bar{b}_L s_R)$ | $4.8 \times 10^2$ | $8.3 \times 10^2$ | $8.8 \times 10^{-6}$ | $2.9 \times 10^{-6}$ | $\Delta m_{B_s}; S_{\psi\phi}$ |

including also the $D$–$\bar{D}$ system, can be found in Ref. [153] (see also Refs. [154, 155] for previous studies). Some of the results, thus obtained, are reported in Table 2.7 and illustrated in Fig. 2.17.[2]

The main messages of these bounds are the following:

- New physics models with a generic flavor structure ($c_{ij}$ of order 1) at the TeV scale are ruled out. If we want to keep $\Lambda$ in the TeV range, physics beyond the SM must have a highly non-generic flavor structure.
- In the specific case of the $|\Delta F| = 2$ operators in (2.72), in order to keep $\Lambda$ in the TeV range, we must find a symmetry argument such that $|c_{ij}| \lesssim |V_{3i}^* V_{3j}|^2$.

The strong constraining power of $|\Delta F| = 2$ observables is a consequence of their strong suppression within the SM. They are suppressed not only by the typical $1/(4\pi)^2$ factor of loop amplitudes, but also by the GIM mechanism and by the hierarchy of the CKM matrix ($|V_{3i}| \ll 1$, for $i \neq 3$). A similar condition is required by consistency with data on rare decays, although in such case the constraints are slightly less stringent.

---

[2]The bounds from $D$–$\bar{D}$ in Fig. 2.17 are more stringent with respect to the values reported in Table 2.7 being based on more recent data [153].

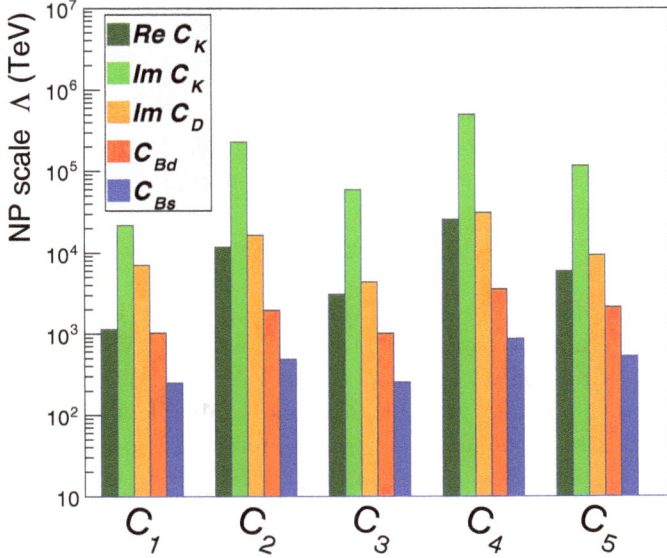

Fig. 2.17: Bounds on the scale of different four-fermion operators contributing to $|\Delta F| = 2$ bounds, assuming an effective coupling $1/\Lambda^2$ (from Ref. [153]), and correspond to the left-handed ($C_1$) and the scalar ($C_4$) operators shown also in Table 2.7, for the definition of the other operators see Ref. [154]. The different colors denote the different type of $|\Delta F| = 2$ transitions used to set the bounds.

Building NP models with new degrees of freedom close to the TeV scale able to satisfy the stringent bounds in Table 2.7 is a highly non-trivial task, which is usually referred to as the *NP flavor problem*. This should be contrasted with the *SM flavor problem*, which is the difficulty of justifying the hierarchical structure of the SM Yukawa couplings pointed out in section 1.1.1.

As already outlined at the end of chapter 1, the way to address the NP flavor problem is to consider the (approximate) flavor symmetry of the SM Lagrangian in the limit of vanishing Yukawa couplings as a good (approximate) symmetry also of the underlying NP model. This generic statement can be made more precise in the context of the effective theory approach to physics beyond the SM by defining an appropriate global flavor symmetry and a set of symmetry breaking terms, as illustrated in appendix B. The most restrictive of such approaches is the so-called Minimal Flavor Violation (MFV) hypothesis [156, 157]. According to this hypothesis, the flavor symmetry of the NP effective Lagrangian is the

maximal global flavor symmetry compatible with the SM gauge group, and
the breaking terms are nothing but the SM Yukawa couplings. However,
this is by no means the only option. For instance, the hypothesis of a
$U(2)^5$ flavor symmetry acting on the light families, minimally broken by
the $3^{\text{rd}}$ generation $\rightarrow$ light generation mixing in the left-handed sector, also
provides a very efficient protection of flavor-changing amplitudes beyond
the SM (see appendix B).

To conclude, we stress that the good agreement of SM and experiments
for $B_{d,s}$ mixing does not imply that further studies of meson mixing are
not relevant. On the one hand, even for $|c_{ij}| \approx |V_{3i}^* V_{3j}|$, which can be
considered the most pessimistic case implied by the MFV hypothesis, we are
presently constraining new physics at the TeV scale. Therefore improving
these bounds, if possible, would be extremely valuable. On the other hand,
as we will discuss in the next chapter, less minimal hypotheses, favored by
the $B$-physics anomalies, suggest a non-trivial pattern of deviations from
the SM in $B_{d,s}$ mixing, quite close to the present limits.

# Chapter 3

# Current anomalies: Experimental evidence

## 3.1 Introduction

In this chapter we discuss the evidences of non-standard effects which in the last few years started to emerge in a series of semileptonic, and purely leptonic, $B$ decays. The evidence appears in three distinct sectors, discussed in the next three sections:

**i.** rare decays of the type $B \to h_s \ell^+ \ell^-$, where $\ell = e, \mu$ and $h_s$ indicates a generic light meson obtained from the initial $B$ meson via a $b \to s$ transition (neutral current);

**ii.** the ultra-rare purely leptonic decay $B_s \to \mu^+ \mu^-$ (helicity-suppressed neutral current);

**iii.** charged-current semileptonic decays of the type $B \to h_c \tau \nu$, where $h_c$ indicates a generic meson obtained via a $b \to c$ transition (charged current).

The system **i** is the richest one in terms of observables, as well as the one exhibiting the most significant discrepancies with respect to the SM. The systems **i** and **ii** are sensitive to the same short-distance amplitude within the SM, but are potentially sensitive to different NP contributions. At first sight, the system **iii** seems to be completely disconnected from **i** and **ii**.

The aspect that connects the three systems is that all the observed anomalies seem to be connected to a possible violation of Lepton Flavor Universality (LFU) of short-distance origin, i.e. a different behaviour of the different lepton species which is not a simple kinematical effect related to their different masses. As we will discuss in the next chapter, the significance of these phenomena, obtained combining all the observations is quite high and is approaching the discovery level. However, before discussing how the different observations combine to obtain a coherent picture, in this chapter we will analyse the three systems separately, describing in detail the various measurements exhibiting deviations from the SM.

## 3.2 The $b \to s\ell^+\ell^-$ anomalies

### 3.2.1 *Generalities*

The first set of anomalies results from studies of the decays $B \to \overline{K}^{(*)}\mu^+\mu^-$, where the notation indicates that either a $\overline{K}$ or $\overline{K}^*(892)$ meson is in the final state. The higher order "penguin" diagrams for these processes are shown in Fig. 3.1. We will discuss measurements of three quantities in several final states: the branching fractions of exclusive $b \to h_s\mu^+\mu^-$ decays, where $h_s$ indicates a hadron, the ratio of branching fractions $\mathcal{B}\left(B \to \overline{K}^{(*)}\mu^+\mu^-\right)/\mathcal{B}\left(B \to \overline{K}^{(*)}e^+e^-\right)$, and the angular distributions in $B \to \overline{K}^*\mu^+\mu^-$, but not $B \to \overline{K}\mu^+\mu^-$, because the spin-0 nature of both the $B$ and the $\overline{K}$ forces the dimuon pair to be in the $(J, J_z) = (1, 0)$ state.

It is necessary to measure the branching fractions of the $b \to h_s\ell^+\ell^-$ decays as a function of the four-momentum transfer squared between the $b$ flavored hadron and the $h_s$, called $q^2$, which can also be evaluated by computing the invariant mass squared of the $\mu^+$ and $\mu^-$. This separation in $q^2$ is necessary, because when the mass of the dimuon pair equals the $J/\psi$ or $\psi(2S)$ masses the diagrams in Fig. 3.1 are swamped, for example, by the reaction $b \to h_s J/\psi$, $J/\psi \to \mu^+\mu^-$, which is depicted in Fig. 1.1(b).[1] In addition, at large $q^2$ it is possible that there is a significant production of

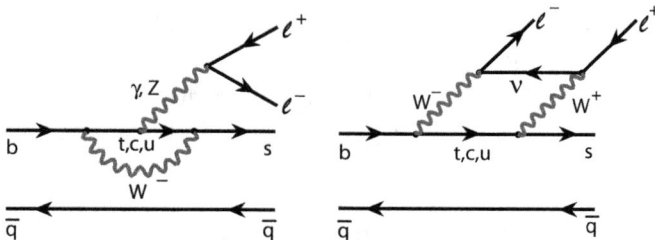

Fig. 3.1: Second order Feynman diagrams for $B \to \overline{K}^{(*)}\mu^+\mu^-$ decays. The $s$ quark and the $\overline{q}$ quark form either a $\overline{K}$ or $\overline{K}^*(890)$ meson.

---

[1] In some cases the region around the $\phi$ mass is also removed due to $\phi \to \mu^+\mu^-$ decays.

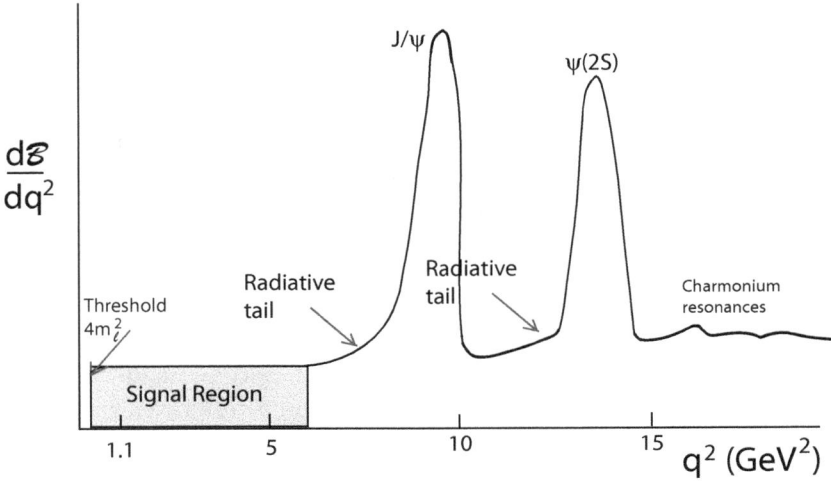

Fig. 3.2: Sketch of the expected $d\mathcal{B}/dq^2$ for $b \to h_s\ell^+\ell^-$ decays. The signal regions for non-resonant decays is typically defined below $\sim$6 GeV$^2$. The $J/\psi$ and $\psi(2S)$ resonances and their radiative tails are also shown. Above the $\psi(2S)$ mass other charmonium resonances are present. If the $h_s$ is a vector particle an additional amplitude with a virtual photon can produce dilepton pairs near threshold, called the "photon pole."

excited charmonium resonances. Indeed, they have already been observed in the $B^+ \to K^+\mu^+\mu^-$ final state [158]. A generalized sketch of the differential branching fraction $d\mathcal{B}/dq^2$ is shown in Fig. 3.2.

### 3.2.2 *Branching fraction measurements of exclusive* $b \to h_s\mu^+\mu^-$ *decays*

First observations of $B \to K^{(*)}\mu^+\mu^-$ decays were made by the Belle collaboration in 2009 [159], the CDF collaboration in 2011 [160], and the BaBar collaboration in 2012 [161]. Unfortunately branching fractions for the dimuon modes either were not quoted, or had large $\sim$25% uncertainties. BaBar and Belle did present averages of the branching fractions for $B \to K^{(*)}\ell^+\ell^-$, where "$\ell$" indicates either electrons or muons. Typically, hadron colliders are far better at detecting muons than electrons, and CDF did not investigate the dielectron mode. The CDF collaboration also made the first observations of $B_s^0 \to \phi\mu^+\mu^-$ [160] and $\Lambda_b^0 \to \Lambda\mu^+\mu^-$ [162]. First attempts were made in these early analyses to investigate angular distributions for NP effects, but the data samples were too small to derive meaningful results.

The first step in more sensitive investigations of phenomena related to $b \to h_s\mu^+\mu^-$ decays was to see if it was possible to observe larger signals, then measure the branching fractions for different $b$ and $h_s$ species, and compare with theoretical predictions. In LHCb analyses electrons are identified in the electromagnetic calorimeter mainly by seeing if there is agreement between the measured track momentum and the energy deposit in the calorimeter, although other criteria are invoked. Muons candidates are required to pass through iron interspersed with multi-layer track detection systems that are sensitive to charged particles. Hadrons are identified in the two Ring Imaging Cherenkov counters. All the LHCb measurements mentioned in this Book use at least one of these identification systems.

Several modes were measured by the LHCb experiment, with data corresponding to either $3\,\mathrm{fb}^{-1}$ of integrated luminosity taken in 7 and 8 TeV $pp$ collisions, called Run 1 or with $\sim 6\,\mathrm{fb}^{-1}$ of Run 2 data taken with 13 TeV $pp$ collisions. In order for the collision data to be recorded "interesting" events need to be selected (triggered upon). LHC experiments have multiple trigger levels, the first one being generated by specific hardware signals. In this case the detection of one or two muons with transverse momentum of the order of $>1.5\,\mathrm{GeV}$ are used. The next trigger levels are software that require there be a collection of tracks including the two muons that is detached from the primary $pp$ collision vertex (PV).

The measured branching fractions can be compared to theoretical model calculations. If they are different then both the experiment and the calculations could be suspect. In general, as stated in chapter 1 the models are based on Light Cone Sum Rules (LCSR) or Lattice QCD (LQCD) calculations. Unfortunately, the more precise LQCD calculations are available only at higher values of $q^2$, while at lower values the predictions are missing, or much less accurate. LHCb measurements in six separate $b \to h_s \mu^+ \mu^-$ exclusive transitions are shown in Fig. 3.3 and compared with theoretical predictions.

In the low $q^2$ region, below $\sim 7\,\mathrm{GeV}^2$, the data are almost all below the predicted branching fractions, while in the high $q^2$ region, the agreement is good, even with the more precise LQCD calculations. For $B_s^0 \to \phi \mu^+ \mu^-$ [175] alone, the measured branching fraction in the $1.1 < q^2 < 6\,\mathrm{GeV}^2$ interval differs from SM prediction by $1.8 - 3.6\,\sigma$ depending on the theoretical model used for the SM prediction [166, 167, 172–174], where the more precise combination of LCSR and LQCD calculations show the larger deviation, and the LCSR calculation alone show the smaller deviation. It is clear however, that the uncertainties on the theory could be correlated among the different modes, so no NP is claimed, but it will become even

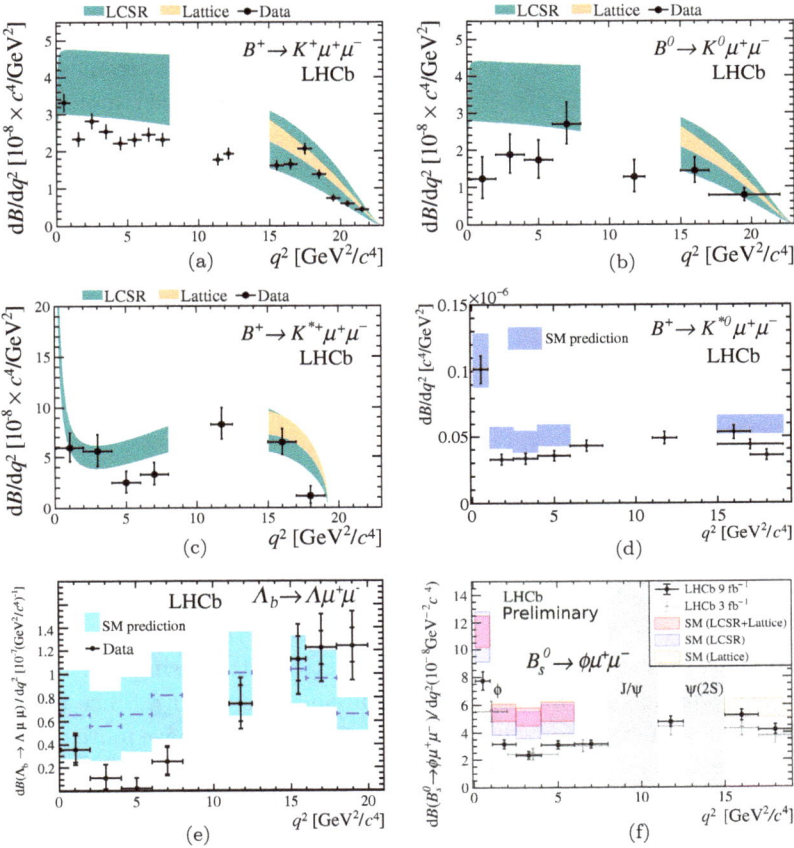

Fig. 3.3: Branching fractions for $b \to h_s \mu^+ \mu^-$ decays compared with theoretical predictions based on the SM. The white or grey vertical bands where experimental points are missing correspond to $q^2$ values equal to the mass-squared of either the $J/\psi$ or $\psi(2S)$ resonances whose widths take into account the experimental mass resolution. For the (a) $B^+ \to K^+ \mu^+ \mu^-$ decays [163], (b) $B^0 \to K^0 \mu^+ \mu^-$ decays [163], and (c) $B^+ \to K^{*+} \mu^+ \mu^-$ decays [163], the SM based predictions are given by [164, 165]. For (d) $B^+ \to K^{*0} \mu^+ \mu^-$ [163] they are given by [166, 167], for (e) $\Lambda_b^0 \to \Lambda \mu^+ \mu^-$ [168] the theory predictions are given by [169], for (f) $B_s^0 \to \phi \mu^+ \mu^-$ [170, 171] they are given by [166, 167, 172–174].

more evident later in this chapter that dimuon final states are below theoretical predictions in the low $q^2$ region. In the cases where $h_s$ is a vector particle, for $q^2$ values close to zero the theoretical predictions all rise substantially due to the additional transition $b \to h_s \gamma$, $\gamma \to \ell^+ \ell^-$. This region is often called "the photon pole."

### 3.2.3   *The LFU ratios in* $B \to K^{(*)}\ell^+\ell^-$ *decays*

As discussed in section 1.4.5, a key property of the SM Lagrangian is Lepton Flavor Universality (LFU). This implies that all charged leptons have identical interactions and that differences in rates, or angular distributions, could only be caused by their masses. These effects can be easily predicted. In fact, differences between electrons and muons in $b$ decays can simply be ignored because their masses are both very small compared to the $b$ quark mass. In 2003 an interesting paper was published by Hiller and Kruger [176] where they showed that changes in the branching fraction ratios

$$R_{K^{(*)}} = \frac{\mathcal{B}\left(B \to \overline{K}^{(*)}\mu^+\mu^-\right)}{\mathcal{B}\left(B \to \overline{K}^{(*)}e^+e^-\right)}, \tag{3.1}$$

of the order of 10% could occur from NP. Recalling that discovery of parity violation, occurred after Lee and Yang deduced that there was no experiment evidence to exclude the effect in weak interactions [177], the LHCb collaboration began to investigate $R_{K^{(*)}}$.

As commented above, having an $e^+e^-$ pair in the final state creates problems for the $b$ hadron detection and reconstruction primarily due to bremsstrahlung radiation from the electrons traversing the detector material, causing their momentum measurement to be low. Since detector requirements generally force hadron collider detectors to have more material than, for example, $e^+e^-$ experiments, they are less able to analyze final states with electrons than muons. A partial remedy is available by using the electromagnetic calorimeter to detect the bremsstrahlung photons. It is not easy, however, as the electrons curve in the magnetic field. Therefore the bremmstrahlung recovery algorithm needs to search for photons that could have been emitted by the charged track along its trajectory before entering the magnetic field. Muons are affected at a much smaller level and also have the advantage of being easier to select in the hardware trigger and reconstruct with excellent momentum resolution.

Although it is possible to directly make the measurement described in Eq. (3.1), the LHCb collaboration decided to measure $R_{K^{(*)}}$ via a double ratio, making use of the $J/\psi$ resonance whose relative decay rate $\Gamma(J/\psi \to e^+e^-)/\Gamma(J/\psi \to \mu^+\mu^-) = 1.0016 \pm 0.0031$, essentially unity. The operative equation is

$$R_{K^{(*)}} = \frac{\mathcal{B}\left(B \to \overline{K}^{(*)}\mu^+\mu^-\right)}{\mathcal{B}\left(B \to \overline{K}^{(*)}e^+e^-\right)} \frac{\mathcal{B}\left(B \to \overline{K}^{(*)}J/\psi;\ J/\psi \to e^+e^-\right)}{\mathcal{B}\left(B \to \overline{K}^{(*)}J/\psi;\ J/\psi \to \mu^+\mu^-\right)}. \tag{3.2}$$

This method mostly cancels many efficiency differences between muons and electrons. Furthermore, the shape of the reconstructed $B$ mass distributions in the $J/\psi$ region can be compared with simulation in both cases and the simulation tuned to be applicable outside of the $J/\psi$ region.

Data in LHCb that is recorded must satisfy both hardware and software requirements (often called "triggers") in LHCb Run 1 and Run 2 data.[2] Hardware requirements are the logical "or" of various selections including identified muons, electrons and large transverse momentum hadrons. Events can be triggered either by looking for specific aspects of a decay, e.g. one or two muons in $B \to K\mu^+\mu^-$ or by the decay of the other $b$-hadron in the event. The software triggers are split into two levels, one for removal of most of the uninteresting events, and the second for more fine tuned selections. Software requirements are varied, but usually include the presence of a decaying long lived hadron, such as a $B$ meson.

Signal reconstruction for all of these modes proceeds by using a neural network rather than making individual selections. This method generally is more efficient and provides more background reduction. The network is trained using simulated signal samples and background samples that are above the known $b$-hadron mass. Several variables are used for input. These are based on the $b$-hadron decay point being significantly displaced from the primary $pp$ interaction vertex (PV), the $\chi^2$ of kaon-dilepton vertex, each decay track being consistent with not coming from the PV and similar topological criteria. Each decay mode has a separate selection, as do different segments of the data taking.

### 3.2.3.1 $R_K$

We start with the LHCb $R_K$ measurement. It was originally made with $3\,\text{fb}^{-1}$ of integrated luminosity [178], was updated using $5\,\text{fb}^{-1}$ [179], and updated using the full $9\,\text{fb}^{-1}$ data sample [180]. Recall that we are interested in measuring $R_K$ in the low $q^2$ region. The high side of the range is determined by the "leakage" of the mis-measured $J/\psi$ particles and is set at $6\,\text{GeV}^2$. The lower limit is caused mainly by contamination in the dielectron mode from converted photons in the detector material and to eliminate other backgrounds such as $B^+ \to \phi K^+$, $\phi \to \mu^+\mu^-$, so $q^2 > 1.1\,\text{GeV}^2$ is used.

The measurement described in Eq. (3.2) requires the determination of four experimental yields and knowledge of their relative efficiencies.

---

[2]The hardware trigger is eliminated in the upgraded detector.

The $K^+\mu^+\mu^-$ candidates are triggered by one of the muons having sufficiently large enough $p_T$ ranging from $1.5 - 1.8\,\text{GeV}$, depending on the data taking time. For $K^+e^+e^-$ events, where yields are lower, there are three possible triggers: a high $p_T$ electromagnetic calorimeter deposit of $2.5 - 3.0\,\text{GeV}$, or a high transverse energy kaon $> 3.5\,\text{GeV}$, or other particles in the event, possibly from the other $b$ decay. The hardware triggered events then have to pass the software trigger requirement of having a vertex detached from the PV.

Let us first discuss the measurement of the yield for the $B^+ \rightarrow K^+J/\psi$; $J/\psi \rightarrow e^+e^-$ channel. Electrons in the LHCb detector radiate bremsstrahlung photons when they cross material. Tracking is separated into stations before and after the large dipole bending magnet. The overall track fit is used to determine the electron momentum, while a match between the track momentum and the energy deposited in the electromagnetic calorimeter is used to distinguish electrons from hadrons. Photons radiated before the magnet affect the measured electron momenta, but they can be found in the electromagnetic calorimeter and added to the corresponding measured momenta. Radiation after the magnet goes into the same calorimeter cells as the electron and it is not necessary to add their energy to the measured electron momentum. These data with large numbers of signal events and small backgrounds are used to verify the simulations of the signal shapes that are used in the low $q^2$ interval. Only photons with transverse energy greater than $75\,\text{MeV}$ are used to add to the track based electron momentum measurement. (The bremsstrahlung recovery is also used in the signal region.)

The candidate $K^+J/\psi$; $J/\psi \rightarrow e^+e^-$ mass spectrum, after applying the neural network selection, is shown in Fig. 3.4 (bottom-left). The signal peak has a approximate Gaussian core and rather large non-Gaussian tails. On the high mass side these are caused by assigning too much energy from radiated photons, while on the low side the radiated photons were not found. The dielectron mass is constrained to be that of the $J/\psi$ meson, and the direction of the $K^+J/\psi$ momentum vertex is constrained to point to the primary vertex. (This is also true for the $J/\psi$ dimuon channel.) There is a potentially a large source of background from $B^+ \rightarrow \overline{D}^0e^+\nu$; $\overline{D}^0 \rightarrow K^+\pi^-$ events, where the $\pi^-$ is wrongly identified as an $e^-$. The branching fraction is a relatively large 2.4% compared to the $4.3 \times 10^{-7}$ signal rate [163] and some of the $\overline{D}^0$ mesons decay without separating significantly from the $B^+$ decay point. These are removed by calculating the $K^+\pi^-$ invariant mass assuming the $e^-$ (or $\mu$) candidate is a $\pi^-$. The background mass distribution

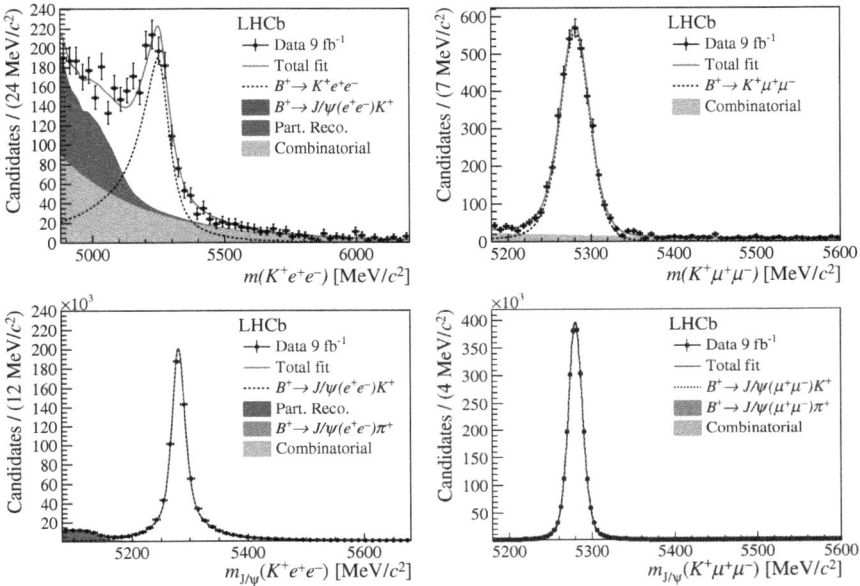

Fig. 3.4: Reconstructed mass distributions for $1.1 < q^2 < 6.0\,\mathrm{GeV}$ (top), $K^+e^+e^-$ (left) and $K^+\mu^+\mu^-$ (right). For dilepton masses consistent with $J/\psi$ decays (bottom), $K^+e^+e^-$ (left) and $K^+\mu^+\mu^-$ (right); here the two leptons are constrained to the $J/\psi$ mass. The dark blue/grey shading shows partially reconstructed background and the light orange/grey the combinatorial backgrounds. Some background components are not visible in the $J/\psi$ sample because they are so small. The solid curves show the total fit, while the dotted curves show the signal components; from Ref. [180].

peaks at the $D^0$ mass; and are easily removed with a cut around the $D^0$ mass. There are also "partially reconstructed" backgrounds that arise from there being a $\overline{K}^*$ in the final state, where the $\pi$ is missed. The combinatorial background comes from mixtures of the partial decay products of two $b$ hadrons. The backgrounds are small. The case where the $J/\psi \rightarrow \mu^+\mu^-$ shown in Fig. 3.4(bottom-right) has more signal events, smaller tails in the signal peak, and much less background.

The mass distributions in the $1.1 < q^2 < 6.0$ interval are shown in Fig. 3.4(top-left) for dielectrons and (top-right) for dimuons. There are approximately $1/500$ fewer events than in the $J/\psi$ resonance region. The data are fit with signal shapes derived from simulation, that are checked with the $J/\psi$ data, and background shapes which also come from simulation but are checked against independent data samples. Simulations that are used to determine efficiencies are tuned on subsamples $K^+J/\psi$ data.

The generated $\overline{B}^0$ $p$, $p_{\mathrm{T}}$, and distributions of associated tracks are weighted to match the observed distributions. Shapes based on the $J/\psi$ mass spectra are used for the dielectron and dimuon channels. Trigger efficiencies are evaluated in data comparing the yields of the different lines [181].

Using Eq. (3.2) the LHCb collaboration determines

$$R_K = 0.846^{+0.042+0.013}_{-0.039-0.012}, \tag{3.3}$$

where the first uncertainty is statistical and the second systematic. The small systematic uncertainty is due to the use of the double ratio to limit the effects of efficiency errors. The result differs from the SM prediction of unity by 3.1 standard deviations and thus hints at an important NP effect.

Two independent checks are done to verify the result. One is just to measure the relative branching fraction of

$$\frac{\mathcal{B}\left(B^+ \to K^+ J/\psi; \ J/\psi \to \mu^+\mu^-\right)}{\mathcal{B}\left(B^+ \to K^+ J/\psi; \ J/\psi \to e^+e^-\right)} = 0.981 \pm 0.020, \tag{3.4}$$

where the uncertainty is both statistical and systematic related to the $R_K$ measurement, and the other is do perform the same measurement using $\psi(2S)$ resonance

$$\frac{\mathcal{B}\left(B^+ \to K^+ \psi(2S); \ \psi(2S) \to \mu^+\mu^-\right)}{\mathcal{B}\left(B^+ \to K^+ \psi(2S); \ \psi(2S) \to e^+e^-\right)} = 0.997 \pm 0.011, \tag{3.5}$$

with the same uncertainties as in the $J/\psi$ ratio measurement. These checks do provide confidence in the result.

The Belle collaboration tested LFU in two reactions $B^+ \to K^+\ell^+\ell^-$ and $B^0 \to K_S\ell^+\ell^-$ [182]. They used their full data sample containing $772 \times 10^6$ $B\overline{B}$ meson pairs. This paper supersedes their earlier results [183], which we will discuss later because it has additional information on the dilepton angular distributions.

Backgrounds and small branching fractions cause potential problems in the analysis. Background arising from $B$ decay usually originate from two uncorrelated leptons in the final state. A possible source is a semileptonic $B$ decay into a charmed hadron that also decays into a lepton, another is an event where both $B$'s decay semileptonically. To suppress these backgrounds and those from continuum $e^+e^-$ collisions, a neutral network is employed.

To measure the yields in each channel a simultaneous fit is done in three variables: (1) the beam constrained mass, $M_{\mathrm{bc}} = \sqrt{E_{\mathrm{beam}}^2 - (\sum_{i=1}^{3} \vec{p_i})^2}$, where $E_{\mathrm{beam}}$ is the energy of the beam in the center-of-mass frame and

Table 3.1: Belle results from the fits for $R_K$, $R_{K_S}$, and their average. Adapted from Ref. [182].

| $q^2$ (GeV$^2$) | Mode | $N_{\mathrm{sig}}$ | $\mathcal{B}$ $(10^{-7})$ | $R_K$ (individual) | $R_K$ (combined) |
|---|---|---|---|---|---|
| | $K^+\mu^+\mu^-$ | $28.4^{+6.6}_{-5.9}$ | $1.76^{+0.41}_{-0.37}\pm0.04$ | $R_{K^+}=$ | |
| $(0.1,4.0)$ | $K^0_S\mu^+\mu^-$ | $6.8^{+3.3}_{-2.6}$ | $0.62^{+0.30}_{-0.23}\pm0.02$ | $0.98^{+0.29}_{-0.26}\pm0.02$ | $1.01^{+0.28}_{-0.25}\pm0.02$ |
| | $K^+e^+e^-$ | $41.5^{+7.7}_{-7.0}$ | $1.80^{+0.33}_{-0.30}\pm0.05$ | $R_{K^0_S}=$ | |
| | $K^0_Se^+e^-$ | $5.5^{+3.6}_{-2.7}$ | $0.38^{+0.25}_{-0.19}\pm0.01$ | $1.62^{+1.31}_{-1.01}\pm0.02$ | |
| | $K^+\mu^+\mu^-$ | $28.4^{+6.4}_{-5.7}$ | $1.24^{+0.28}_{-0.25}\pm0.03$ | $R_{K^+}=$ | |
| $(4.00,8.12)$ | $K^0_S\mu^+\mu^-$ | $4.2^{+4.2}_{-3.5}$ | $0.27^{+0.18}_{-0.13}\pm0.01$ | $1.29^{+0.44}_{-0.39}\pm0.02$ | $0.85^{+0.30}_{-0.24}\pm0.01$ |
| | $K^+e^+e^-$ | $26.9^{+6.9}_{-6.1}$ | $0.96^{+0.24}_{-0.22}\pm0.03$ | $R_{K^0_S}=$ | |
| | $K^0_Se^+e^-$ | $9.3^{+3.7}_{-3.0}$ | $0.52^{+0.21}_{-0.17}\pm0.02$ | $0.51^{+0.41}_{-0.31}\pm0.01$ | |
| | $K^+\mu^+\mu^-$ | $42.3^{+7.6}_{-6.9}$ | $2.30^{+0.41}_{-0.38}\pm0.05$ | $R_{K^+}=$ | |
| $(1.0,6.0)$ | $K^0_S\mu^+\mu^-$ | $3.9^{+2.7}_{-2.0}$ | $0.31^{+0.22}_{-0.16}\pm0.01$ | $1.39^{+0.36}_{-0.33}\pm0.02$ | $1.03^{+0.28}_{-0.24}\pm0.01$ |
| | $K^+e^+e^-$ | $41.7^{+8.0}_{-7.2}$ | $1.66^{+0.32}_{-0.29}\pm0.04$ | $R_{K^0_S}=$ | |
| | $K^0_Se^+e^-$ | $8.9^{+4.0}_{-3.2}$ | $0.56^{+0.25}_{-0.20}\pm0.02$ | $0.55^{+0.46}_{-0.34}\pm0.01$ | |

$\vec{p_i}$ is the measured three-momentum in the center-of-mass for the kaon and the two leptons, (2) the difference in the measured energy and the beam energy, and (3) the output of the neural net designed to reduce backgrounds. The results are shown in Table 3.1. (The projections of the fits in the low $q^2$ region are not publicly available.)

A comparison of the Belle and LHCb results is shown in Fig. 3.5. Clearly the Belle results are not accurate enough to distinguish between the SM and the LHCb $R_K$ value. An older BaBar result is shown as well.

### 3.2.3.2 $R_{K^*}$

The latest measurement using the $K^{*0}(890)$ resonance referred to in Eq. (3.2) used LHCb data corresponding to $3\,\mathrm{fb}^{-1}$ in 7 and 8 TeV $pp$ collisions and $2\,\mathrm{fb}^{-1}$ from 13 TeV collisions [184]. The hardware trigger strategy is the same as that for $R_K$, described in section 3.2.3.1. There are two signal region $q^2$ intervals chosen, one at $0.45 < q^2 < 1.1\,\mathrm{GeV}^2$ and the other at $1.1 < q^2 < 6.0\,\mathrm{GeV}^2$. The lower boundary of the first interval is close to dimuon threshold. The boundary between the two intervals is chosen so that any contamination from $\phi \to K^+K^-$ decays is contained

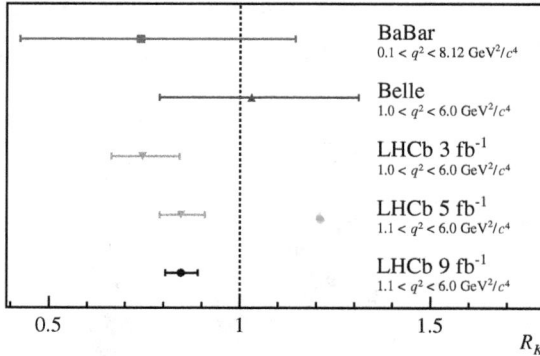

Fig. 3.5: $R_K$ results in the low $q^2$ region compared with Belle and BaBar. The $9\,\text{fb}^{-1}$ LHCb measurement supersedes their previous results.

in the lower interval, which is sensitive to different NP effects than the larger one. The upper end of the second interval is chosen to exclude $K^{*0}J/\psi$ decays. For electrons bremsstrahlung recovery is used as described in section 3.2.3.1.

Events are selected based on a neutral network that uses as input how well the four tracks form a common vertex, the impact parameter of the $B^0$ with respect to the PV, the flight distance of the $B^0$, the consistence of the $B^0$ momentum vector and its line of flight. There are other variables input to the net; perhaps the most useful is $p_T$ balance along the $B^0$ direction of the $K^+\pi^-$ with the dielectron pair (not used for dimuons).

There are specific backgrounds that need special treatment. In the $J/\psi \to \mu^+\mu^-$ mode the resolution is good enough that reconstructed $K^+\pi^-J/\psi$ events with masses below $5150\,\text{MeV}$ are obvious backgrounds. This provides a normalization in this $q^2$ region for simulations. The same is true for $J/\psi \to e^+e^-$, but the two electrons must be constrained to the $J/\psi$ mass in order to provide adequate mass resolution. The decay rate for $B^0 \to D^-\ell^+\nu$, $D^- \to K^{*0}\ell^-\overline{\nu}$ is $10^4$ times larger than the signal. Since there are two missing neutrinos the reconstructed mass does not peak near the $B^0$ mass. However if the neutrinos are both low momentum this decay could populate the signal region, especially for electrons with the poorer resolution. This background, however, is eliminated by examining the opening angle between the two leptons.

Sometimes a muon from the $J/\psi$ decay can be switched in identity with the kaon or pion. These events are eliminated by calculating the mass of the hadron candidate (using the muon mass) with the oppositely charged muon and insuring that is not consistent with being a $J/\psi$ meson. Another potential background is formed by real $K^+\ell^+\ell^-$ events that could pick up a random pion and fake signal. Such decays are vetoed by requiring signal candidates to have a $K^+\ell^+\ell^-$ mass $< 5100\,\mathrm{MeV}$. Finally, background from $B_s^0 \to \phi\ell^+\ell^-$ decays is eliminated by ensuring that the invariant mass of the $\pi^-$, when interpreted as a $K^-$ and combined with the $K^+$, is above the $\phi$ mass.

In order to determine the number of signal events the $K^+\pi^-\ell^+\ell^-$ events in each of the three $q^2$ intervals are first fit constraining them to come from a single vertex and to point at the origin [185]. This procedure improves the mass resolution. The fits are done separately for each trigger category and the three electron categories are summed in Fig. 3.6. Shape parameters other than the mean and width for the signal functions are shared among the electron modes and separately for the muon modes. For the $J/\psi$ $q^2$ region the mass fits are shown in $(a)$ and $(b)$. The signals in both the $e^+e^-$ and $\mu^+\mu^-$ channels are dominant with just small backgrounds from misidentified $\overline{\Lambda}_b^0 \to K^+\bar{p}J/\psi$ events where the $\bar{p}$ is identified as a $\pi^-$ and the Cabibbo suppressed process $\overline{B}_s^0 \to K^{*0}J/\psi$. The combinatorial background is too small to be seen. The $K^{*0}\mu^+\mu^-$ final states shown in $(d)$ and $(f)$ are very clean with only small combinatorial backgrounds. Most of the background in the $K^{*0}e^+e^-$ data in $(c)$ comes from events with an $e^+e^-$ pair where other tracks fake the $K^{*0}$, while in the larger $q^2$ region shown in $(e)$ there are additional combinatorial backgrounds and predicted background from leakage of $J/\psi$ events.

Efficiencies are determined either from data, e.g. the trigger, or from simulation which has been matched to other data. The results for $R_{K^*}$ using the double ratio as well as the single ratio $R_{J/\psi}$ are shown in Table 3.2 and Fig. 3.7. For $R_{K^*}$ the systematic uncertainties are smaller than the statistical ones. $R_{J/\psi}$ is determined to be consistent with unity with a systematic uncertainty of 4.5%; part of this error is assigned to $R_{K^*}$, along with mass fit shapes, trigger, PID, etc.

The results are compared with theoretical predictions in Fig. 3.7. The theoretical predictions agree very well with each other, as a consequence of the good theory control on LFU ratios (see chapter 4). The predictions are smaller than unity in the lowest $q^2$ bin due to the presence of the "photon pole" (See section 3.2.1.) Depending on which theory is used the results are $2.1 - 2.3\sigma$ below the prediction for the lower $q^2$ bin and $2.4 - 2.5\sigma$ in the central $q^2$ bin.

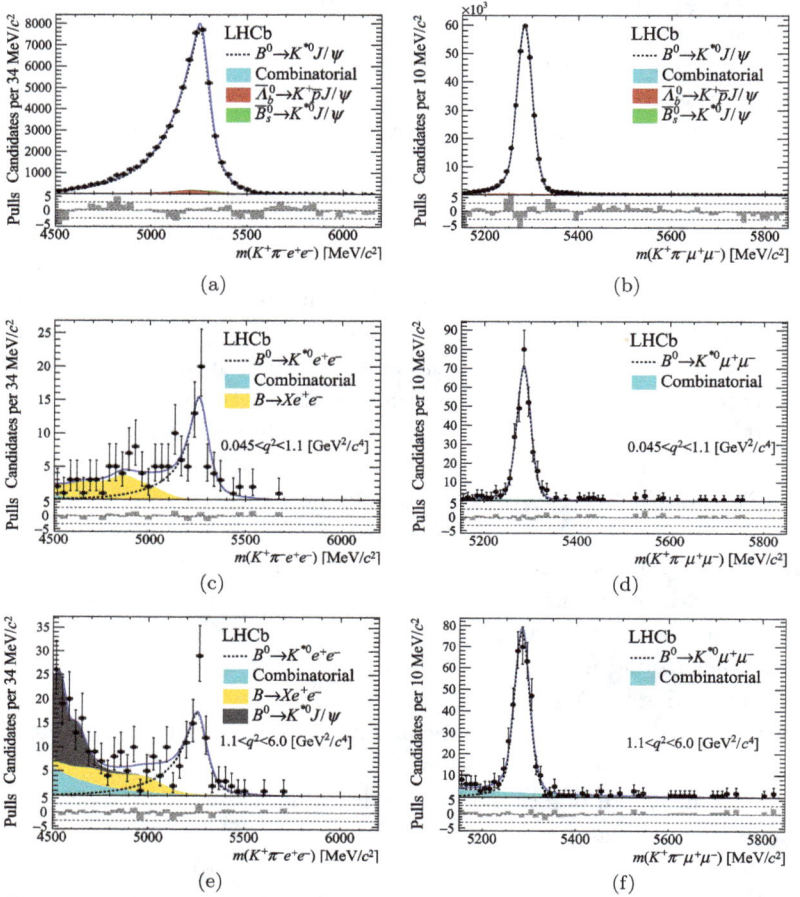

Fig. 3.6: For $K^+\pi^-$ masses within $\pm 100\,\mathrm{MeV}$ of the $K^{*0}(890)$ mass, the results of the fits for different final states are shown: (a) Invariant mass of $K^+\pi^- e^+ e^-$ and (b) $K^+\pi^-\mu^+\mu^-$ candidates when the dilepton mass is in the $J/\psi$ resonance region. (c) Invariant mass of $K^+\pi^- e^+ e^-$ and (d), $m(K^+\pi^-\mu^+\mu^-)$, for $1.1 < q^2 < 6.0\,\mathrm{GeV}$. (e) Invariant mass of $K^+\pi^- e^+ e^-$ and (f), $m(K^+\pi^-\mu^+\mu^-)$, for $1.1 < q^2 < 6.0\,\mathrm{GeV}$. The data are described by black points with error bars, the partially reconstructed backgrounds are shaded in teal, the combinatorial background shaded in light orange. Very small backgrounds in (a) and (b) are shown shaded in red and green, The total fit to the signal and background shapes is shown by the solid blue curves. The signal is shown with dotted lines. Adapted from [184].

The Belle collaboration has analyzed $B^0 \to K^{*0}\ell^+\ell^-$ decays, where $K^{*0} \to K^+\pi^-$, and $B^+ \to K^{*+}\ell^+\ell^-$ decays where $K^{*+} \to K_S\pi^+$ or $K^+\pi^0$ [186]. The average of the $B^0$ and $B^+$ results is shown in Fig. 3.8. These results will improve in accuracy with the addition of Belle II data.

Table 3.2: Results of the $R_{K*}$ measurements using the double ratio, as well as the single ratio for $J/\psi \rightarrow \mu^+\mu^- J/\psi \rightarrow e^+e^-$.

| Ratio | Value | $q^2$ interval $(\text{GeV}^2)$ |
|---|---|---|
| $R_{K*}$ | $0.66^{+0.11}_{-0.07} \pm 0.03$ | $0.45 < q^2 < 1.1\,\text{GeV}^2$ |
| $R_{K*}$ | $0.69^{+0.11}_{-0.07} \pm 0.05$ | $1.1 < q^2 < 6.0\,\text{GeV}^2$ |
| $R_{J/\psi}$ | $1.043 \pm 0.006 \pm 0.045$ | near $m^2_{J/\psi}$ |

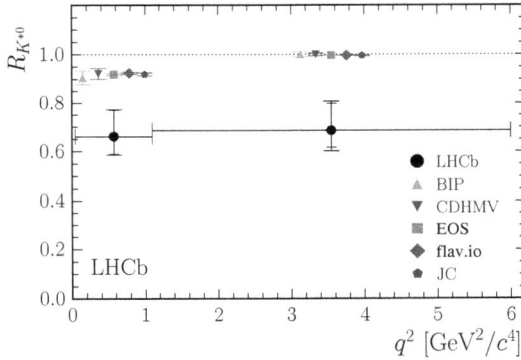

Fig. 3.7: The measured values of $R_{K^{0*}}$ shown as circles with two error bars; the first one shows the statistical uncertainty, while the second the effect of adding the systematic uncertainty. The theoretical predictions are from BIP [187], CDHMV [188–190], EOS [191], flav.io [167, 192], and JC [193]. Data from [184].

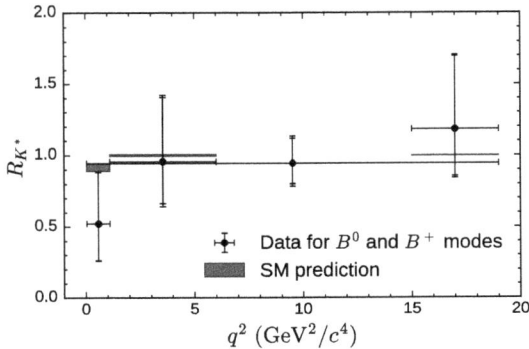

Fig. 3.8: $R_{K*}$ averaged over $B^0$ and $B^+$ modes from Belle [186] compared with the SM predictions from [189, 195].

**Other LFU ratios**

The LHCb collaboration has also analyzed $\Lambda_b^0 \to pK^-\ell^+\ell^-$ decays [194] using $4.7\,\text{fb}^{-1}$ of 7, 8 and $13\,\text{TeV}$ $pp$ collision data, and found a deficit in the ratio of muons to electrons. The ratio is $0.86^{+0.14}_{-0.11} \pm 0.05$ in the range $0.1 < q^2 < 6.0\,\text{GeV}$, showing a $\sim 1\sigma$ deficit consistent with other LHCb measurements. Clearly, more data is needed in this mode.

### 3.2.4  *Angular distributions in $B \to K^* \mu^+ \mu^-$*

Deviations of decay angular distributions from SM predictions can be indicative of NP. In $B \to K\mu^+\mu^-$ the initial $B$ meson is spin-0 as is the $K$ forcing the vector like dimuon pair to be polarized in the $(J, J_z) = (1, 0)$ state. Thus, study of the angular distributions in this mode can be used to confirm the understanding of the detector acceptance, but not to search for NP. There is more freedom in the $B \to K^* \mu^+ \mu^-$ decay as the meson in the final state has spin one. The analysis is done summing over $B^0$ and $\overline{B}^0$ final states, the distinguishing characteristic being the sign of the kaon charge, $K^+$ for $B^0$ and $K^-$ for $\overline{B}^0$. Most of the results from the ATLAS [196], Belle [183], CMS [197] and LHCb [198] collaborations come from analysis of the $B^0 \to K^{*0}\mu^+\mu^-$; $K^{*0} \to K^+\pi^-$ reaction, with some results from LHCb [199] and Belle [183] for the $B^+ \to K^{*+}\mu^+\mu^-$; $K^{*+} \to K_S\pi^+$ channel. The Belle results also contain a test of lepton universality, so we will discuss them separately later.

We first give a detailed description of angular distributions, then discuss the experimental results, and compare with theoretical predictions. As in any other decay of a spinless hadron into a vector-vector final state the decay distribution is a function of three angles. Consider the $B^0$ decay: (1) The angle between the $\mu^+(\mu^-)$ and the direction opposite to that of the $B^0$ ($\overline{B}^0$) in the rest frame of the dimuon system labeled by $\theta_\ell$. (2) The angle between the direction of the $K^+(K^-)$ and the $B^0$ ($\overline{B}^0$) in the rest frame of the $K^{*0}(\overline{K}^{*0})$ system is denoted as $\theta_K$. (3) The angle between the plane defined by the dimuon pair and the plane defined by the kaon and pion in the $B^0$ ($\overline{B}^0$) rest frame is denoted by $\phi$. These angles are shown in Fig. 3.9. (For the $B^\pm$ decays the definitions are similar.)

The decay amplitude $\Gamma$ for the $B^0$ decay where the $K$ and $\pi$ are in a $P$-wave, as it would be for a pure $K^*(892)$ resonance, can be decomposed in full generality as [200, 201]:

$$\frac{d^4\Gamma}{dq^2\,d\cos\theta_K\,d\cos\theta_l\,d\phi} = \frac{9}{32\pi}J(q^2, \theta_K, \theta_l, \phi) \qquad (3.6)$$

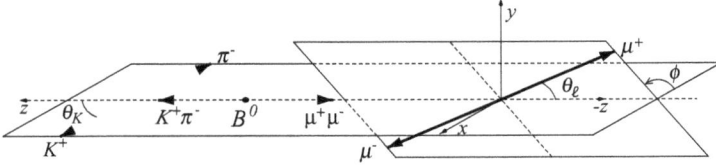

Fig. 3.9: Pictorial representation of the decay angles. See text for definitions.

$$J(q^2, \theta_K, \theta_l, \phi)$$
$$= \left[ J_{1s} \sin^2 \theta_K + J_{1c} \cos^2 \theta_K + (J_{2s} \sin^2 \theta_K + J_{2c} \cos^2 \theta_K) \cos 2\theta_l \right.$$
$$+ J_3 \sin^2 \theta_K \sin^2 \theta_l \cos 2\phi + J_4 \sin 2\theta_K \sin 2\theta_l \cos \phi$$
$$+ J_5 \sin 2\theta_K \sin \theta_l \cos \phi + (J_{6s} \sin^2 \theta_K + J_{6c} \cos^2 \theta_K) \cos \theta_l$$
$$+ J_7 \sin 2\theta_K \sin \theta_l \sin \phi + J_8 \sin 2\theta_K \sin 2\theta_l \sin \phi$$
$$\left. + J_9 \sin^2 \theta_K \sin^2 \theta_l \sin 2\phi \right]. \tag{3.7}$$

The $J_i$'s are all functions of $q^2$. They are related to the transversity amplitudes in the $B \to K^*$ transition. They can be calculated in the SM or other models but this incurs theoretical uncertainties. We are interested here in the average of $B$ and $\overline{B}$. Other measurements can be made of the $CP$ difference amplitudes and will be very interesting with more statistics. To average the $B$ and $\overline{B}$ rates, we take advantage of the fact that the decay rate $\overline{\Gamma}$ of the $CP$-conjugated process $\overline{B^0} \to \overline{K}^*(\to K\pi)\ell^+\ell^-$ is obtained from Eq. (3.6) by replacing $J_{1,2,3,4,7} \to \bar{J}_{1,2,3,4,7}$ and $J_{5,6,8,9} \to -\bar{J}_{5,6,8,9}$, where $\bar{J}$ is equal to $J$ with all weak phases conjugated. This corresponds to the same definition of $\theta_\ell$ for both $B$ and $\overline{B}$ (see for example [202, 203]). Here all the observables are $CP$-averaged, and so are always functions of $J_i + \bar{J}_i$. Therefore, $J_i \to J_i + \bar{J}_i$ and $\Gamma \to \Gamma + \overline{\Gamma}$ should be understood in all the formulas below, and in particular all the $J$ observables are taken to be $(J + \bar{J})$.

The next step is to simplify matters by introducing folded distributions. The symmetry $\phi \leftrightarrow \phi + \pi$ has been used to produce a "folded" angle $\hat{\phi} \in [0, \pi]$, so that $d\hat{\Gamma}(\hat{\phi}) = d\Gamma(\phi) + d\Gamma(\phi - \pi)$. The differential decay rate becomes [204]

$$\frac{d^4\Gamma}{dq^2 \, d\cos\theta_K \, d\cos\theta_l \, d\hat{\phi}} = \frac{9}{16\pi} J(q^2, \theta_K, \theta_l, \hat{\phi}). \tag{3.8}$$

This expression can be rewritten in terms of other measurable quantities that are easier to interpret. The shape of the decay amplitude $\Gamma$, the average of $B^0$ and $\bar{B}^0$ decay widths for the $K$ and $\pi$ in a $P$-wave, as it would be for a pure $K^*$ resonance, defining $d\vec{\Omega} \equiv d\cos\theta_K d\cos\theta_l d\hat{\phi}$, is given by [203, 205]:

$$\frac{1}{d(\Gamma)/dq^2} \frac{d^4(\Gamma)}{dq^2 \, d\vec{\Omega}}\bigg|_P = \frac{9}{32\pi} \bigg[ \frac{3}{4}(1 - F_L)\sin^2\theta_K + F_L\cos^2\theta_K$$

$$+ \frac{1}{4}(1 - F_L)\sin^2\theta_K \cos 2\theta_l$$

$$- F_L\cos^2\theta_K \cos 2\theta_l + S_3 \sin^2\theta_K \sin^2\theta_l \cos 2\phi$$

$$+ S_4 \sin 2\theta_K \sin 2\theta_l \cos\phi + S_5 \sin 2\theta_K \sin\theta_l \cos\phi$$

$$+ \frac{4}{3}A_{FB}\sin^2\theta_K \cos\theta_l + S_7 \sin 2\theta_K \sin\theta_l \sin\phi$$

$$+ S_8 \sin 2\theta_K \sin 2\theta_l \sin\phi + S_9 \sin^2\theta_K \sin^2\theta_l \sin 2\phi \bigg],$$

$$(3.9)$$

where $F_L$ is the fraction of the longitudinal polarization of the $K^{*0}$ meson, $A_{FB}$ is the forward-backward asymmetry of the dimuon system and $S_i$ are other $q^2$ dependent $CP$-averaged observables [205]. Some of parameters, e.g. $F_L$ and $A_{FB}$ were considered important to measure because they could reveal NP. However, these parameters come with somewhat large theoretical uncertainties as they involve predictions of hadron transition form-factors.

Theoretical work showed that other $q^2$ dependent combinations of the form-factors in Eq. (3.9) had significantly smaller uncertainties [201, 206]. Generally experiments use these definitions [196–198, 205][3]

$$P_1 = \frac{2S_3}{(1 - F_L)} = A_T^{(2)} \quad P_2 = \frac{2}{3}\frac{A_{FB}}{(1 - F_L)} \quad P_3 = \frac{-S_9}{(1 - F_L)}$$

$$P'_{4,5,8} = \frac{S_{4,5,8}}{\sqrt{F_L(1 - F_L)}} \quad P'_6 = \frac{S_7}{\sqrt{F_L(1 - F_L)}} \,.$$

$$(3.10)$$

In addition to the resonant P-wave $K^{*0}$ contribution to the $K^+\pi^-\mu^+\mu^-$ final state, the $K^+\pi^-$ system can also be in an S-wave configuration. S-waves in the presence of larger P-waves were introduced in Ref. [135]. The addition of an S-wave component introduces two new complex amplitudes

---

[3]The definition of the $P'_i$ observables differs somewhat from that of Ref. [201].

and results in six additional angular terms. The angular distribution is modified as

$$
\frac{1}{d(\Gamma)/dq^2} \frac{d^4(\Gamma)}{dq^2\, d\vec{\Omega}}\bigg|_{S+P} = (1 - F_S) \frac{1}{d(\Gamma)/dq^2} \frac{d^4(\Gamma)}{dq^2\, d\vec{\Omega}}\bigg|_P
$$

$$
+ \frac{3}{16\pi} F_S \sin^2\theta_l
$$

$$
+ \frac{9}{32\pi}(S_{11} + S_{13}\cos 2\theta_l)\cos\theta_K
$$

$$
+ \frac{9}{32\pi}(S_{14}\sin 2\theta_l + S_{15}\sin\theta_l)\sin\theta_K\cos\phi
$$

$$
+ \frac{9}{32\pi}(S_{16}\sin\theta_l + S_{17}\sin 2\theta_l)\sin\theta_K\sin\phi,
$$

(3.11)

where $F_S$ denotes the S-wave fraction, and the terms $S_{11}$, $S_{13}$–$S_{17}$ represent the interference between the S and P-wave amplitudes. Note that $F_S$ replaces the terms $S_{10}$ and $S_{12}$, with $F_S = 3S_{10} = -3S_{12}$. In the analyses we will now discuss, these S-wave terms are used only by the LHCb [198] and CMS [197] collaborations, while the ATLAS [196] and BELLE [183] collaborations omit them. It is likely that the level of accuracy needed for the results to be influenced by the S-wave terms has not yet been reached in those experiments.

We first discuss the LHC experiments. LHCb results for the $B^0 \to K^{*0}\mu^+\mu^-$, $K^{*0} \to K^+\pi^-$ channel [198] are based on data collected with $3\,\mathrm{fb}^{-1}$ of luminosity at 7 and 8 TeV, and $1.7\,\mathrm{fb}^{-1}$ at 13 TeV. As the $b$-hadron cross-section doubles from 7 to 13 TeV, the results presented here represent roughly twice the statistics present in the previous publication [205]. The LHCb results for the $B^+ \to K^{*+}\mu^+\mu^-$, $K^{*+} \to K_S\pi^+$ channel [199] are based on the full LHCb sample of $9\,\mathrm{fb}^{-1}$. ATLAS [196] and CMS [197] use data collected at 8 TeV from a little over $20\,\mathrm{fb}^{-1}$ of luminosity. While it looks that ATLAS and CMS would have larger statistics than LHCb in these rare modes, other restrictions, such as the trigger, actually reverse that impression. All three experiments use different trigger strategies, although dimuon hardware triggers are part of the mix. The requirements on the $p_T$ of the muons are quite different. CMS requires each muon have $p_T >$ 3.5 GeV and be in the pseudorapidity range $|\eta| < 2.2$, while in the same $\eta$ range ATLAS requires the $p_T$ of one muon be $>4$ GeV and the other 6 GeV. ATLAS also has many other trigger lines that have various criteria.

LHCb typically requires the muons to have $p_T > 1.5$ GeV. Other LHCb selections are applied in the software trigger. For further analysis selections, all three experiments have $p_T$ selections on the kaon and pion from the $K^{*0}$ and requirements on the four final state tracks to form a vertex detached from the PV. LHCb does a signal to background optimization using a neural network (BDT).

Further restrictions are placed on the $K^+\pi^-\mu^+\mu^-$ final state. Since $\phi$, $J/\psi$, and $\psi(2S)$ resonances can decay into $\mu^+\mu^-$ data with $q^2$ consistent with their masses within a few experimental widths are removed. In addition the $K^+\pi^-$ invariant mass is restricted to be close to the $K^*(890)$ mass, typically within $\pm100$ MeV. LHCb has the luxury of particle identification so that the kaon and pion are identified, while ATLAS and CMS must consider both combinations, chose one, and then simulate how often they made the wrong choice in order to correct the final results for the mis-identification.

Using the angular distribution measurements for each event, the fitted values of the coefficients defined in Eqs. (3.9), (3.10), and (3.11) can be determined and compared with theoretical predictions as functions of $q^2$. This requires good knowledge of the angular acceptances. They are determined by a combination of simulation and checks using $B^0 \rightarrow J/\psi K^{*0}$ decays, where the polarization amplitudes have been well measured [207]. Due to the contributions of resonant $\phi \rightarrow \mu^+\mu^-$, $J/\psi \rightarrow \mu^+\mu^-$, and $\psi(2S) \rightarrow \mu^+\mu^-$ decays, the LHCb measurements leave out the $q^2$ intervals around $q^2$ equal to the mass squared of these particles. ATLAS does not remove the $\phi$ contribution and CMS includes a small part of it. (The size of the intervals depends on the mass resolution of each experiment.) The $K^+\pi^-\mu^+\mu^-$ mass distributions are shown in Fig. 3.10 for selected low $q^2$ intervals for the three experiments. The reader can clearly see the differences in signal yields, background fractions, and mass resolutions for the different experiments. Only about half the yields are shown for the LHCb experiment, the rest are in the 13 TeV data. In (c) and (d) the CMS experiment has nicely shown the effect of kaon-pion mis-tagging. ATLAS accounts for it by having a mis-tagging rate that they apply in the analysis. The particle identification in LHCb has eliminated this component.

The experiments differ in their mass resolutions, signal yields, and background fractions, all of which vary with $q^2$. The mass resolution and background fraction for the LHCb $K_S\pi^+\mu^+\mu^-$ analysis (not shown) is similar to the $K^+\mu^+\mu^-$ analysis. The number of signal events quoted by the experiments is summarized in Table 3.3.

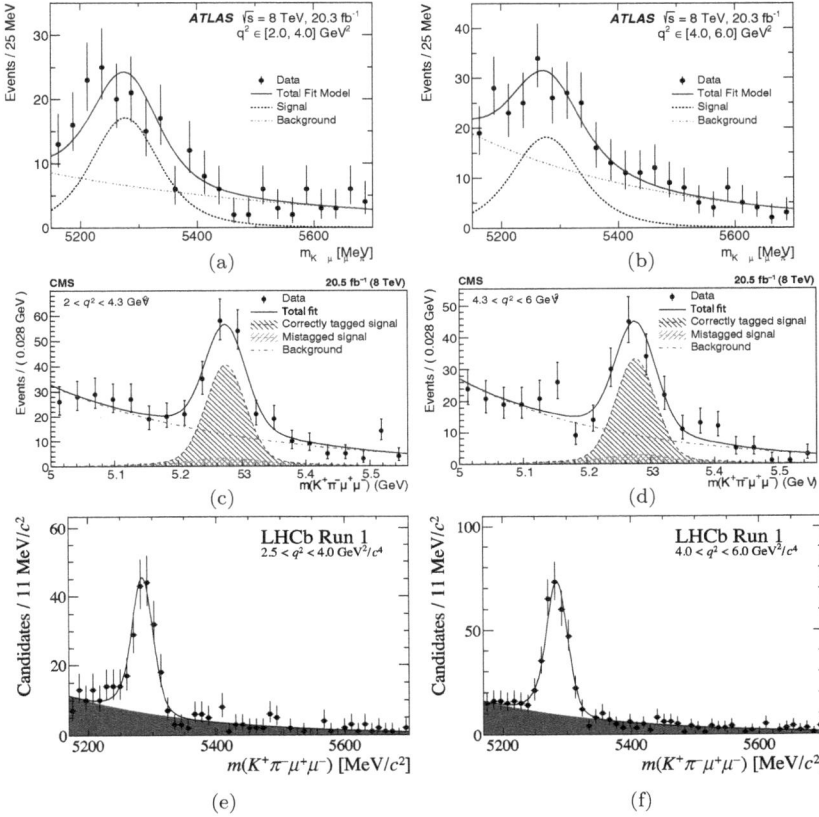

Fig. 3.10: Reconstructed $K^+\pi^-\mu^+\mu^-$ mass distributions for some $q^2$ intervals below 6 GeV$^2$ from (a)(b) ATLAS, (c)(d) CMS and (e)(f) LHCb. The LHCb plots are shown only for 7 and 8 TeV data; there is approximately an equal number of events in 13 TeV data. The background is blue shaded in (e) and (f). (Please note the different mass bin sizes.)

Table 3.3: Ranges of $q^2$ where measurements in $K^*\mu^+\mu^-$ modes were made and signal event yields in these intervals, from different experiments. Results from the Belle experiment will be discussed later.

| Exp. | $q^2$ interval (GeV$^2$) | Signal yields |
|------|--------------------------|---------------|
| ATLAS | $0.04 < q^2 < 6.0$ | $342 \pm 39$ |
| CMS | $1.0 < q^2 < 19.0$ | $1397 \pm 45$ |
| LHCb $B^0$ | $1.1 < q^2 < 19.0$ | $4585 \pm 78$ |
| LHCb $B^+$ | $1.1 < q^2 < 19.0$ | $737 \pm 34$ |
| Belle $B^0 + B^+$ | $0.1 < q^2 < 19.0$ | $185 \pm 17$ |

Some experiments report only a subset of the $q^2$ dependent angular distributions in Eqs. (3.9), (3.10), and (3.11), while LHCb presents all of the distributions in the supplemental material of the paper [198]. These are then compared with SM calculations, in particular "DHMV" [188, 208], "ASZB" [166, 167], and "JC" [193, 209]. The LHCb collaboration uses the FLAVIO software package [172] to fit the angular variables in the $q^2$ bins below $q^2$ of 8 GeV$^2$, and one bin in the interval $15 < q^2 < 19$ GeV$^2$, by varying the Wilson coefficient $C_9$. The best fit is obtained with shift with respect to the SM of $\mathrm{Re}(C_9) = -0.99^{+0.25}_{-0.21}$, corresponding to a significance of $3.3\sigma$. In this analysis other Wilson coefficients have not been allowed to differ from their SM values. This is because the modification in $\mathrm{Re}(C_9)$ already provides a very good fit, and this modification could also be responsible for the LFU anomalies. We will discuss in more detail SM predictions and BSM interpretations after the presentation of the experimental results. Some of the LHCb $B^0$ results are shown in Fig. 3.11.

Most of the LHCb results are in reasonable agreement with these predictions, except for the comparison with $P_5'$ where two bins are above

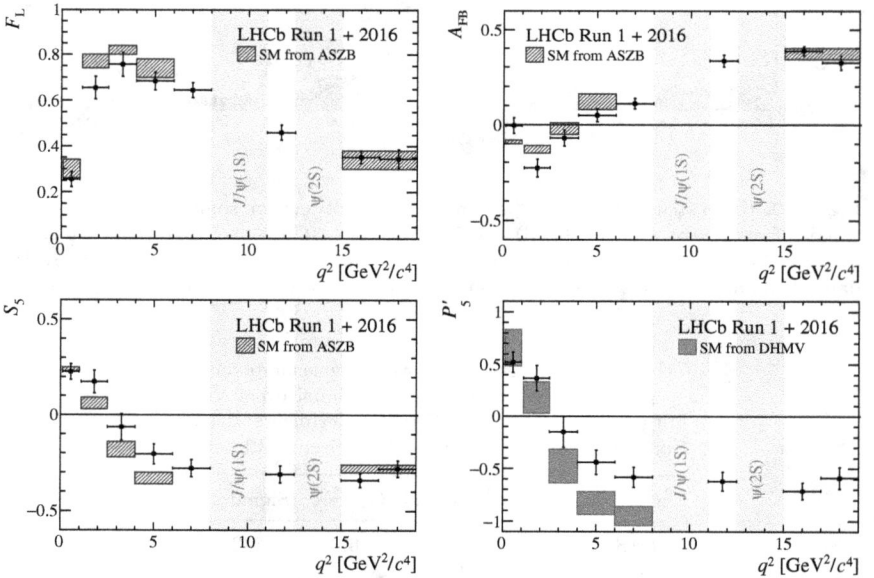

Fig. 3.11: Some angular coefficients for $B^0 \to K^{*0}\mu^+\mu^-$ decays as functions of $q^2$ measured by LHCb [198] and compared with SM calculations either from ASZB [166, 167] or DHVM [188, 208]. Systematic uncertainties are included in the error bars.

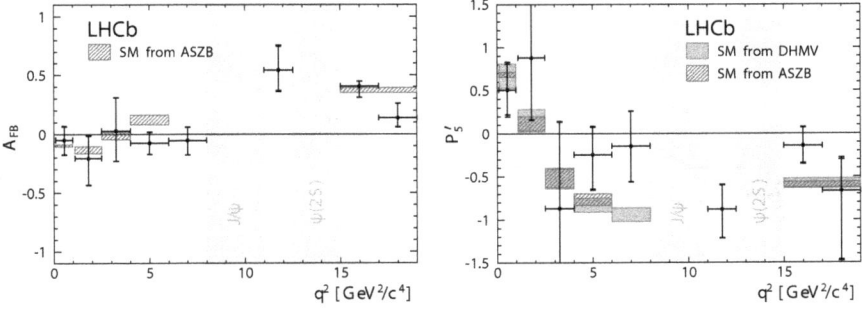

Fig. 3.12: $A_{FB}$ and $P'_5$ for $B^+ \to K^{*+}\mu^+\mu^-$ decays as functions of $q^2$ measured by LHCb and compared with SM calculations either from ASZB [166, 167] or DHVM [188, 208]. There are two vertical error bars on the data points; the separation is hard to see. The inner one shows the statistical uncertainly and the outer one the combination in quadrature of the statistical and systematic uncertainties. The vertical grey columns show the $q^2$ excluded regions around the $\phi$, $J/\psi$, and $\psi(2S)$ squared masses.

the SM prediction. The local discrepancies are 2.5 standard deviations ($\sigma$) in the bin $4 < q^2 < 6 \,\mathrm{GeV}^2$, and $2.9\sigma$ in the bin $< 6 < q^2 < 8 \,\mathrm{GeV}^2$. Considering all the variables and all the bins examined, it appears these result are not too significant at this time. However, as we will see, they can be combined with other anomalies to draw some consequences. The results of the smaller statistics LHCb $B^+$ analysis are shown in Fig. 3.12. While the data are above the SM predictions in the $4 < q^2 < 8 \,\mathrm{GeV}^2$ region, the uncertainties are too large to draw any conclusions.

Some of the ATLAS results are shown in Fig. 3.13. In the $P'_5$ comparison the data are higher than the SM predictions in one bin: $4 < q^2 < 6 \,\mathrm{GeV}^2$. Unfortunately, there are no data shown for the higher $8 > q^2 > 6 \,\mathrm{GeV}^2$ interval.

Next, in Fig. 3.14, we view some of the CMS results. Their findings are consistent with the SM and the other experimental results.

We have one more result to view from the Belle collaboration [183], before we summarize this section, that measures $P'_5$ separately for $K^*\mu^+\mu^-$ and $K^*e^+e^-$ final states, thus providing both a test of NP in the deviation with the SM and the difference in dimuon and dielectron final states. The $B^0 \to K^{*0}\ell^+\ell^-$ and $B^+ \to K^{*+}\ell^+\ell^-$ channels are investigated with the full data sample corresponding 711 fb$^{-1}$ of integrated luminosity taken at the peak of the $\Upsilon(4S)$ resonance. The data are summed over $B^0$ and $B^+$ candidates. The $K\pi$ mass is restricted to be between 0.6 and 1.4 GeV. The analysis uses a neural network to optimize signal with respect to

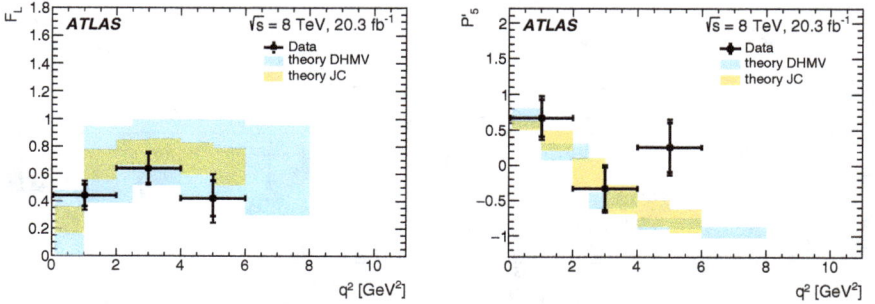

Fig. 3.13: $F_L$ and $P_5'$ as functions of $q^2$ measured by ATLAS for $B^0 \to K^{*0}\mu^+\mu^-$ decays and compared with SM calculations either from DHVM [188, 208] or JC [193, 209]. The inner error bar on the data shows the statistical uncertainty, while the outer one has the systematic uncertainty added in quadrature.

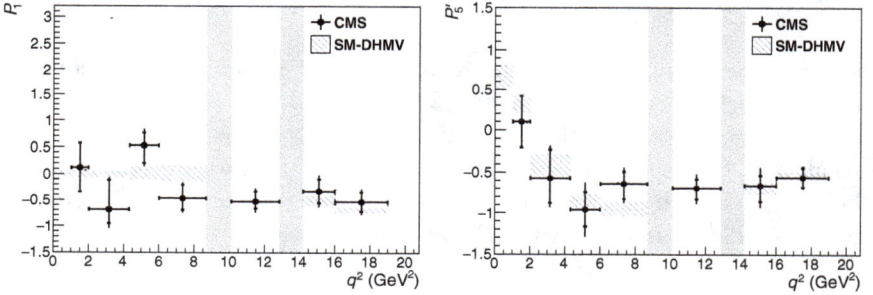

Fig. 3.14: $P_1$ and $P_5'$ as functions of $q^2$ for $B^0 \to K^{*0}\mu^+\mu^-$ decays measured by CMS and compared with SM calculations from DHVM [188, 208]. The dash on the error bar on the data shows the statistical uncertainty, while the extended line has the systematic uncertainty added in quadrature.

background. They follow the normal procedure at $e^+e^-$ colliders of viewing the invariant mass of the $K\pi\ell^+\ell^-$ after replacing the measured $B$ candidate energy with half the center-of-mass energy, equal to the $B$ energy ($E_B$) and calling it the "beam constrained mass" ($M_{bc}$), as described in Eq. (3.12).

$$M_{bc}^2 = (E_{K^*} + E_{\mu^+} + E_{\mu^-})^2 - \left(\sum \vec{P}_{K^*} + \vec{P}_{\mu^+} + \vec{P}_{\mu^-}\right)^2$$

$$= (E_B)^2 - \left(\sum \vec{P}_{K^*} + \vec{P}_{\mu^+} + \vec{P}_{\mu^-}\right)^2. \tag{3.12}$$

The resulting $K\pi e^+e^-$ and $K\pi\mu^+\mu^-$ candidate mass spectrum are shown in Fig. 3.15. The distributions are shown over all $q^2$ with regions near the

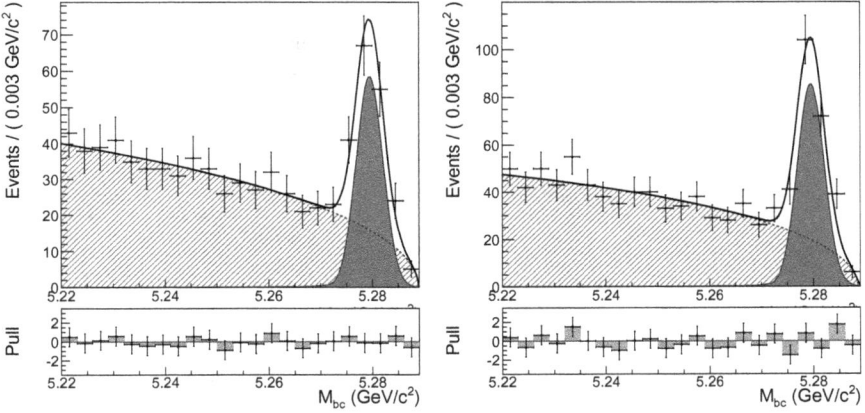

Fig. 3.15: $M_{bc}$ for (left) $K\pi e^+e^-$ and (right) $K\pi\mu^+\mu^-$ candidates from Belle [183], fit with a Crystal Ball function for signal (dark grey) [210], and an ARGUS shape for combinatorial background (light grey) [211].

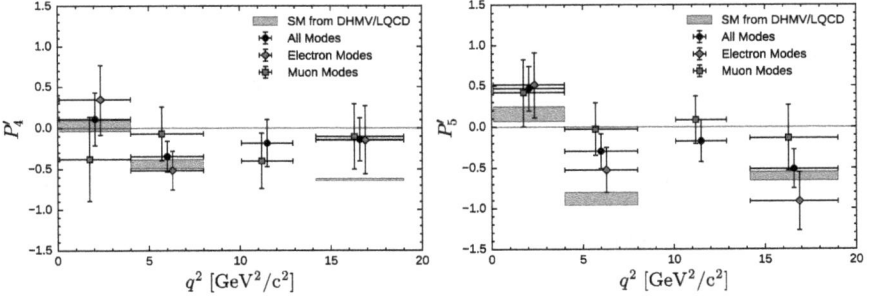

Fig. 3.16: Measured distributions of (left) $P_4'$ and (right) $P_5'$ from Belle [183] for both $K^*e^+e^-$ (dark circles), $K^*\mu^+\mu^-$ (red circles), and their average (black circles). The SM theory from DHVM [188, 208] is also shown (grey rectangles).

$J/\psi$ and $\psi(2S)$ masses removed. There are $127 \pm 15$ and $185 \pm 17$ signal events over a substantial background.

The results of the determination of the angular coefficients $P_4'$ and $P_5'$ are shown in Fig. 3.16 for both dielectron and dimuon final states. The limited event numbers do not allow for definitive conclusions. While the $P_4'$ data is consistent with the SM prediction, in the $P_5'$ $q^2$ bin from 4–8 GeV$^2$, the dimuon data is $2.3\sigma$ above the SM prediction, and somewhat higher than the dielectron data.

Table 3.4: $P_5'$ values in ranges of $q^2$ for $B \to K^* \mu^+ \mu^-$ measurements and their local deviations, data minus the SM prediction in terms of $\sigma$. The CMS data has been corrected slightly to allow it to be included even though the bin edges do not match. Averages (1) and (2) do not include the Belle data because they are not separated into two $q^2$ bins. Where two uncertainties are quoted, the first is statistical and the second systematic. The SM predictions are from [188].

| Exp. | Particle | $q^2$ interval (GeV$^2$) | $P_5'$ | #$\sigma$ deviation |
|------|----------|------------------------|--------|---------------------|
| SM | | $4.0 < q^2 < 6.0$ | $-0.82 \pm 0.08$ | |
| SM | | $6.0 < q^2 < 8.0$ | $-0.94 \pm 0.08$ | |
| ATLAS | $B^0$ | $4.0 < q^2 < 6.0$ | $0.26 \pm 0.35 \pm 0.18$ | 2.5 |
| Belle | $B^0 + B^+$ | $4.0 < q^2 < 8.0$ | $-0.03^{+0.31}_{-0.30} \pm 0.09$ | 2.1 |
| CMS | $B^0$ | $4.3 < q^2 < 6.0$ | $-0.96^{+0.22}_{-0.21} \pm 0.25$ | $-0.5$ |
| CMS | $B^0$ | $6.0 < q^2 < 8.7$ | $-0.64^{+0.15}_{-0.19} \pm 0.13$ | 0.9 |
| LHCb | $B^0$ | $4.0 < q^2 < 6.0$ | $-0.44 \pm 0.11 \pm 0.11$ | 2.5 |
| LHCb | $B^0$ | $6.0 < q^2 < 8.0$ | $-0.58 \pm 0.09 \pm 0.03$ | 2.7 |
| LHCb | $B^+$ | $4.0 < q^2 < 6.0$ | $-0.25^{+0.32}_{-0.40} \pm 0.09$ | 1.3 |
| LHCb | $B^+$ | $6.0 < q^2 < 8.0$ | $-0.15^{+0.40}_{-0.41} \pm 0.06$ | 1.6 |
| Avg. (1) | $B^0 + B^+$ | $4.0 < q^2 < 6.0$ | $-0.41 \pm 0.11$ | 3.2 |
| Avg. (2) | $B^0 + B^+$ | $6.0 < q^2 < 8.0$ | $-0.60 \pm 0.12$ | 2.1 |

To show the various contributions of the experiments to hints at discrepancies with the SM in the $P_5'$ variable, the numerical results in the low $q^2$ bins are listed for the $B \to K^* \mu^+ \mu^-$ modes in Table 3.4. We combine the local deviations in Averages (1) and (2) along with the Belle data to find the local significance of the deviations in the $4.0 < q^2 < 8.0 \, (\text{GeV}^2)$ range. Here we average the statistical and systematic uncertainties in quadrature, but keep the systematic uncertainties for each experiment as common and not averaged in the two bins. We find a local significance of $4.4\sigma$. However, as tempting as it is to get excited by the significance of this result we must caution the reader that the "look elsewhere" effect must be considered. Here what we mean by looking elsewhere arises from the fact that there are 7 variables to be measured, as listed in Eq. (3.10). Each one has many bins to examine. If we, for example, only consider those below $q^2 < 8 \, \text{GeV}^2$, then we have a total of 5 bins to consider for each variable. How often will two contiguous bins both be above say $2\sigma$? Considering two contiguous bins out of 5 in each variable gives 4 combinations, and multiplying by 7 yields 28 possibilities. Given a Gaussian distribution of measurements the probability of exceeding $2\sigma$ is 4.6% in each bin, since both positive and negative fluctuations are allowed. So the probability of finding two contiguous bins outside of $|2\sigma|$ is 6%. Furthermore, the theoretical modeling has to be treated with a great deal of caution. In fact, as can be seen in Fig. 3.13, the JC model has a somewhat large central value and a larger uncertainty than DHMV, which reduces the size of the significances. This is not the end of the story, however, as these considerations only apply to the initial LHCb result. Other experiments then concentrated on these particular intervals and, therefore, do not have a look elsewhere effect. In conclusion, there is tantalizing, but not conclusive evidence of NP in these angular distributions. More sophisticated analyses have fit all the angular distributions simultaneously both for SM and allowing variations of the Wilson coefficients, thus allowing for non-SM behavior. These results, thus far, have significances below $5\sigma$ [198], but avoid concentrating on specific bins with local deviations.

### 3.2.5 *Summary of $B \to K^{(*)}\ell^+\ell^-$ decays*

LHCb has observed an evidence, at the $3.1\sigma$ level of significance, of a violation of Lepton Flavor Universality in $B \to K^+\ell^+\ell^-$ decays at low $q^2$: muons turn out to be suppressed in rate with respect to electrons by 15.4%. Using less data, the $\overline{B} \to K^{*0}\ell^+\ell^-$ decays show a similar but larger $\sim$30% suppression in each of the two $q^2$ regions below 6 GeV$^2$. In generic extensions of the SM there is no reason for these two suppression factors to be the same; however, they are expected to be very close in motivated NP frameworks (see chapter 4). While it is not clear how systematically correlated the LHCb measurements are, in all the three cases systematic errors are much smaller than the statistical ones. Considering the LFU measurements as fully uncorrelated, the three measurements of $R_{K^{(*)}}$ represent a deviation from the SM hypothesis somewhat above $4\sigma$.

The angular analyses, and the rate measurements, can point to the same underlying NP responsible for the deviations from the SM in $R_{K^{(*)}}$. However, in this chapter we refrain from including the $P_5'$ deviations in an overall estimate of the significance of NP, for two main reasons: i) there is no complete consensus about the theoretical error in the SM predictions in $P_5'$, with sizeable variations among different groups; ii) the most significant discrepancies have been in a particular region of $q^2$, and the quoted discrepancies with the theoretical predictions are local significances. An estimate of the global significance of LFU and angular observables necessarily requires a more detailed analysis of the theoretical predictions. A conservative estimate of the global significance of the NP hypothesis,

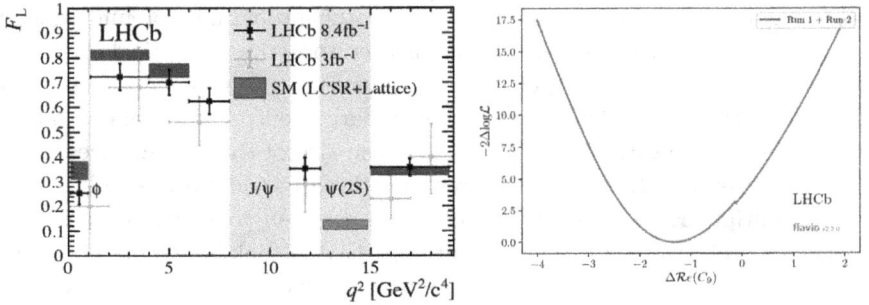

Fig. 3.17: (left) Distribution of the $F_L$ angular variable as a function of $q^2$ from LHCb data compared with the theory derived from LSCR plus input from LQCD [167, 172–174]. (right) Scan of −2 times the logarithmic liklihood for the SMEFT parameter $\Delta\mathcal{R}e(\mathcal{C}_9)$.

taking into account the effect of there being other $q^2$ bins and other angular variables that have been examined, and the effect that different NP hypotheses lead to different correlations, is presented in chapter 4.

A consistent global analysis of all the angular variables in Eq. (3.9), but not the LFU ratios, has been performed by LHCb [198]. This analysis shows a disagreement with the SM hypothesis of about $3\sigma$ or less, depending on the choice of the $q^2$ bins.

### 3.2.6   *Angular analysis of $B_s^0 \to \phi\mu^+\mu^-$ decay*

These decays are quite similar to $B^0 \to K^{*0}\mu^+\mu^-$ with the angular variables having the same definitions. Here there is less angular information than in the $K^*$ case because $\phi\mu^+\mu^-$ doesn't tag the final state as coming from a $B_s^0$ or $\overline{B}_s^0$, so they are summed. The LHCb collaboration has done a full angular analysis for the $B_s^0 \to \phi\mu^+\mu^-$ decay [212]. It shows a small deviation of $-1.9\sigma$ in the $\Delta\mathcal{R}e(\mathcal{C}_9)$ SMEFT coefficient, mostly due a discrepancy in the $F_L$ distribution shown in Fig. 3.17. This is quite consistent, if not with the $F_L$ distribution shown in Fig. 3.11 for $B^0 \to K^{*0}\mu^+\mu^-$ decay, although by itself it is statistically insignificant.

### 3.3   The decays $B_s^0 \to \mu^+\mu^-$ and $B^0 \to \mu^+\mu^-$

Another decay that proceeds via a second order weak interactions in the SM is $B_s^0 \to \mu^+\mu^-$. The diagrams are shown in Fig. 3.18. These decays cannot be compared with the corresponding $B_s^0 \to e^+e^-$ and $B^0 \to e^+e^-$ decays because of the "helicity suppression" present in these specific transitions. The reason of this suppression can be understood as follows: $B$ mesons

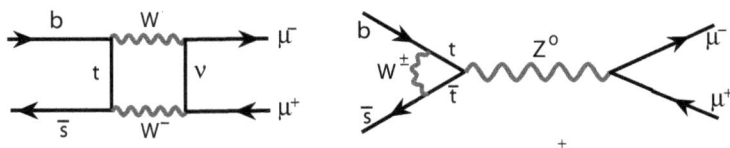

Fig. 3.18: Second order Feynman diagrams for $B_s^0 \to \mu^+\mu^-$ decays.

are spin-0 objects. Consider the viewpoint of the $B$ rest-frame. If a $B$ decays into a spin-1/2 fermion and a spin-1/2 anti-fermion, the spins must be opposite to conserve angular momentum. However, the fermion is left-handed and the anti-fermion right-handed, so the natural tendency is for the spins to be aligned. In fact, the decay rate would be zero for massless fermions, as it is proportional to $(m_\ell/m_B)^2$, thus suppressing the $e^+e^-$ with respect to the $\mu^+\mu^-$ final states by a factor of 42,787.

Since the $B \to \mu^+\mu^-$ amplitudes are highly suppressed in the SM, but still measurable, NP can interfere with the SM diagrams and substantially change the decay rates. Thus, we compare the measured branching fractions with the SM predictions of [213]

$$\mathcal{B}(B_s^0 \to \mu^+\mu^-) = (3.66 \pm 0.14) \times 10^{-9} \, ,$$
$$\mathcal{B}(B^0 \to \mu^+\mu^-) = (1.03 \pm 0.05) \times 10^{-10} \, , \qquad (3.13)$$
$$\frac{\mathcal{B}(B^0 \to \mu^+\mu^-)}{\mathcal{B}(B_s^0 \to \mu^+\mu^-)} = (2.81 \pm 0.16)\% \, .$$

Model builders usually calculate the partial width of the decay, in this case $\Gamma(B \to \mu^+\mu^-)$, and then divide by the total width ($\Gamma$) to predict the branching fractions. The final $\mu^+\mu^-$ final state is $CP$-odd, which is associated with the "heavy" $B$ decay width $\Gamma^H$, as opposed to the $CP$-even "light" decay width. This is important for $B_s^0$ decays because the measured time integrated branching fractions can be compared directly with these predictions [214], as the authors did use $\Gamma_s^H$ to compute the branching fractions. (For $B^0$ decays, the decay width difference is negligible.) It is also possible to start measuring the effective lifetime ($\tau_s^{\text{eff}}$) and comparing with current measurements of $\tau_s^H = 1/\Gamma_s^H$.

The analysis for all three experiments is complicated by the small branching fractions and backgrounds including two-body $B$ meson decays into final states such as $\pi^+\pi^-$, $K^+K^-$, and $K^\pm\pi^\mp$, and the semileptonic decays $\pi^-(K^-)\mu^+\nu$. A neural network framework, a Boosted Decision Tree (BDT), is used by all three experiments, although different variables

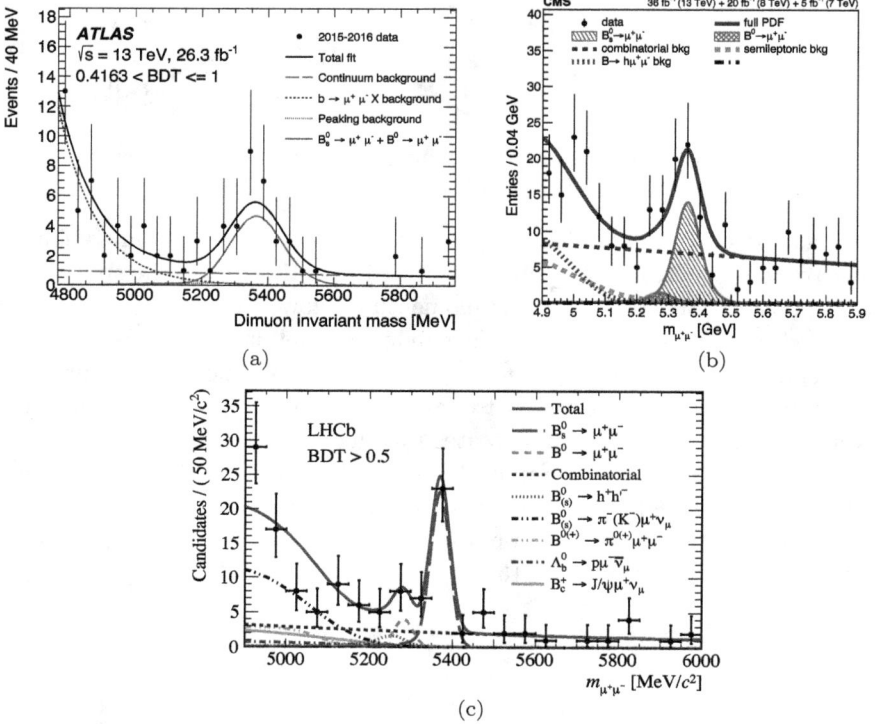

Fig. 3.19: The $\mu^+\mu^-$ mass spectra for (a) ATLAS [215] (b) CMS [216], and (c) LHCb [217] experiments. Fitted background components are shown along with the $B^0$ and $B_s^0$ candidate signal shapes. Only the 13 TeV ATLAS data is shown. The remainder can be viewed in Ref. [218].

are used in each of them. This is necessary because of different detector capabilities, especially the particle identification capability of LHCb. The resulting $\mu^+\mu^-$ mass spectra in the most signal like BDT bins are shown for each of the experiments in Fig. 3.19. Robust signals are seen for $B_s^0 \to \mu^+\mu^-$, while the $B^0$ signal is not yet observed. The data are normalized to separate measurements of $B^- \to J/\psi K^-$ decays using the ratio of measured $B_s^0$ to $B^0$ production fractions [219].[4]

The results from three LHC experiments are summarized in Table 3.5, along with computed averages. The lifetime is consistent with the current value of $\tau_s^H = 1.620 \pm 0.007$ ps [12]. LHCb has recently updated their

---

[4]LHCb also uses the $B^0 \to K^+\pi^-$ mode for normalization.

Table 3.5: Values of $\mathcal{B}(B_s^0 \to \mu^+\mu^-)$, $\mathcal{B}(B^0 \to \mu^+\mu^-)$, and $(\tau_s^{\text{eff}})$ from the three different LHCb experiments and the averages. Recent LHCb results not included in the average are also listed.

| Exp. | $\int \mathcal{L}$ (fb$^{-1}$) 7+8, 13 TeV | $\mathcal{B}(B_s^0 \to \mu^+\mu^-)$ $\times 10^{-9}$ | $\mathcal{B}(B^0 \to \mu^+\mu^-)$ $\times 10^{-10}$ | $(\tau_s^{\text{eff}})$ ps |
|---|---|---|---|---|
| ATLAS [215, 218] | 25.0, 26.3 | $\left(2.8^{+0.8}_{-0.7}\right)$ | $(-1.9 \pm 1.6)$ | |
| CMS [216] | 25.0, 36.0 | $\left(2.9^{+0.7}_{-0.6} \pm 0.2\right)$ | $\left(0.8^{+1.4}_{-1.3}\right)$ | $1.70^{+0.60}_{-0.43} \pm 0.09$ |
| LHCb [217] | 3.0, 1.4 | $\left(3.0 \pm 0.6^{+0.3}_{-0.2}\right)$ | $\left(1.5^{+1.2+0.2}_{-1.0-0.1}\right)$ | $2.04 \pm 0.44 \pm 0.5$ |
| Avg. [223] | | $\left(2.69^{+0.37}_{-0.35}\right)$ | $(0.6 \pm 0.7)$ $<1.9$ @ 95% cl. | $1.91^{+0.37}_{-0.35}$ |
| LHCb [220] | 3.0, 6.0 | $\left(3.09^{+0.46+0.15}_{-0.43-0.11}\right)$ | $\left(1.2^{+0.8}_{-0.7} \pm 0.1\right)$ | $2.07 \pm 0.29 \pm 0.03$ |

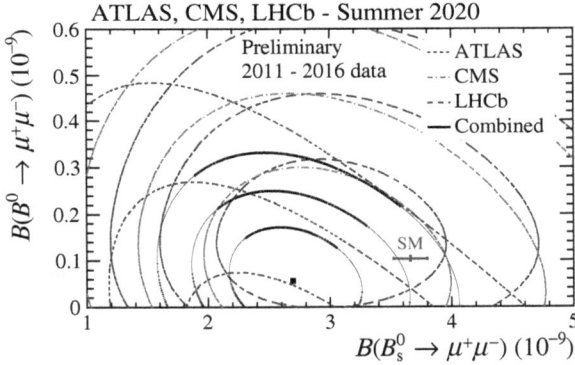

Fig. 3.20: Likelihood contours for $\mathcal{B}(B_s^0 \to \mu^+\mu^-)$ versus $\mathcal{B}(B_s^0 \to \mu^+\mu^-)$ for each experiment and combined [223]. The dashed contours correspond to the individual experiments and the black averaged ones show changes in $-2$ of the natural logarithm of the likelihood of 2.3, 6.2 and 11.8. The square-root of these numbers gives the approximate number of standard deviations of the contours from the measured point. The SM predictions are shown by the red cross.

analysis using the entire Run 1 plus Run 2 data sample [220], also using an updated ratio of $B_s^0$ to $B^0$ production [221]. These results are listed separately as they are not part of the average. The $B^0$ and $B_s^0$ branching fractions are correlated, so we show in Fig. 3.20 the results from each experiment, but only including the LHCb result used in the average, the average, along with the SM predictions. In the two dimension branching fraction plane, the SM prediction for the $B_s$ mode is $2.1\sigma$ above the measured average including the theoretical uncertainty on the prediction. As we discus in chapter 4, this deviation from the SM, which by itself is not very significant, can possibly be connected to the deviations from the SM in $R_{K^{(*)}}$.

## 3.4   Anomalies in the ratio of tauonic to muonic semileptonic decays

### 3.4.1   *Introduction*

The semileptonic decays we discuss here are described to lowest order by the Feynman diagram shown in Fig. 1.1(a). Before discussing the tests of $\tau^-$ to $\mu^-$ lepton flavor universality in semileptonic $B$ decays, we note that the most precise test of electron-muon universality in these decays was performed by the Belle collaboration [224].

$$\frac{\mathcal{B}(B^0 \to D^{*-}e^-\nu)}{\mathcal{B}(B^0 \to D^{*-}\mu^-\nu)} = 1.01 \pm 0.01 \pm 0.03 \ . \tag{3.14}$$

Other tests using $D$ rather than $D^*$ decays are not as stringent [225]. In fact, it is not expected that NP would show up in these tree-level decays as the SM rate is not suppressed. Despite this theoretical prejudice the BaBar collaboration decided to test lepton flavor universality by measuring the ratios $\mathcal{B}\left(\overline{B} \to D^{(*)}\tau^-\overline{\nu}\right)/\mathcal{B}\left(\overline{B} \to D^{(*)}\ell^-\overline{\nu}\right)$ where $\ell$ stands for in this case either $e^-$ or $\mu^-$ [226, 227]. Their results created a great deal of interest. To explain the measurements and their significance we will try and illustrate the different techniques used by the BaBar, Belle and LHCb experiments to make these measurements, and then summarize the results.

Measurements of $\overline{B} \to D^*\ell^-\overline{\nu}$ were made by the ARGUS [228] and CLEO [229] $e^+e^-$ experiments operating at the $\Upsilon(4S)$ resonance mainly in order to determine the CKM element $|V_{cb}|$ [230]. Unlike the analyses described above in chapter 3.1 where all the particles are visible and the signal can be found in $\Upsilon(4S)$ decays using Eq. (3.12), here there is a missing neutrino, and another method must be used to distinguish the signal from background. The analysis proceeds by finding events with candidate $D^*$ and lepton decays. If the three-momentum of the $B$ were known, then the missing mass squared could be computed as

$$\mathrm{MM}^2 = [E_B - (E_{D^*} + E_{\ell^-})]^2 - [\vec{p}_B - (\vec{p}_{D^*} + \vec{p}_{\ell^-})]^2 \ , \tag{3.15}$$

where $E_i$ and $\vec{p}_i$ refer to the energy and three-momentum of particle $i$. The $B$ energy is half the center-of mass energy, or equivalently, the beam energy in the center-of-mass frame ($E_{\text{beam}}$) giving

$$\text{MM}^2 = [E_{\text{beam}} - (E_{D^*} + E_{\ell^-})]^2 - [\vec{p}_B - (\vec{p}_{D^*} + \vec{p}_{\ell^-})]^2. \qquad (3.16)$$

As the direction of $\vec{p}_B$ is unknown, it is set to zero

$$\text{MM}^2 \approx [E_{\text{beam}} - (E_{D^*} + E_{\ell^-})]^2 - [\vec{p}_{D^*} + \vec{p}_{\ell^-}]^2, \qquad (3.17)$$

which widens the experimental resolution. Since the neutrino mass is so small, a peak at zero is expected. Indeed it is found, but the width of $\text{MM}^2$ peak of $0.8 \text{ GeV}^2$ fwhm, is sufficiently large that in also contains backgrounds including those from $B \to D^{**}\ell^- \bar{\nu}$ final states, where $D^{**}$ is a generic name for all excited $D^*\pi$ or $D^*\pi\pi$ states that are produced either resonantly or non-resonantly and constitute a significant and poorly know background. Another set of backgrounds are $B \to D^{*,**}D_s^{-(*)}$, $D_s^- \to \tau^-/\mu^-\bar{\nu}$ decays, although these are much smaller and weren't considered at the time. Data taken at center-of-mass energies just below $B\bar{B}$ threshold were used to estimate backgrounds from $e^+e^- \to q\bar{q}$ processes with $q = u, d, s$ or $c$, called "continuum" processes.

### 3.4.2 *Measurements of $R_{D^{(*)}}$ using fully reconstructed B meson tags at B-factories*

It is possible at $B$-factories to fully reconstruct one $B$ meson by detecting directly all of its decay products (called a $B_{\text{tag}}$) and then find final states of the other $B$ meson in the event with one missing particle. When the missing particle is a neutrino, the gain in doing this is that the $\text{MM}^2$ resolution is improved by about a factor of four, since both the $B$ energy and three-momentum are known in Eq. (3.15). Backgrounds are also severely reduced. The penalty is that actual number of $B$ signal events is drastically reduced. For example, using a "NeuroBayes"-based optimization, tagging efficiencies of $0.28\%$ and $0.18\%$ were achieved for $B^+$ and $B^0$ mesons in Belle data [231]. Clearly this represents a substantial loss of statistical power, on the other hand it does allow some analyses to be done that could not be conducted in any other manner. An example is measurement of $B \to D\ell\nu$ as opposed to $D^*$ decays that was done by Belle [225] and BaBar [232].

To test $\tau^-/\mu^-$ lepton flavor universality in semileptonic $B$ decays, the ratios of both $D$ and $D^*$ final states are subject to measurement

$$R_{D^*} \equiv \frac{\mathcal{B}(\overline{B} \to D^*\tau^-\overline{\nu})}{\mathcal{B}(B \to D^*\ell^-\overline{\nu})}, \quad R_D \equiv \frac{\mathcal{B}(\overline{B} \to D\tau^-\overline{\nu})}{\mathcal{B}(B \to D\ell^-\overline{\nu})}, \tag{3.18}$$

where $D^*$ means either $D^{*+}$ or $D^{*0}$, and $D$ means $D^0$ or $D^+$. These numbers are integrated over the phase space of the decays which can be parameterized in terms of the energy of the lepton and $q^2$, the four-momentum transfer squared between the $B$ and the detected charm hadron.

The Standard Model predictions for $R_{D^{(*)}}$ are not unity because of two considerations: i) the phase space is very different because of the presence of a heavy $\tau^-$ lepton instead of an $e^-$ or $\mu^-$, as shown in Fig. 3.21, and ii) the form-factors describing the $\tau^-$ lepton are more complicated because its mass allows for significant right-handed helicity terms. There have been several theoretical predictions [233–236]. The average of these reported by the Heavy Flavor Averaging Group (HFLAV) [74] is $R_D = 0.299 \pm 0.003$, and $R_{D^*} = 0.258 \pm 0.005$. A recent detailed prediction [237] gives the same value for $R_D$, but has $R_{D^*} = 0.247 \pm 0.005$. This difference is substantial.

Finding a $MM^2$ peak in semileptonic $B$ mesons decays containing $\tau^-$ leptons is not possible because the $\tau^-$ decay has at least one missing neutrino, while the $B$ decay also has one missing neutrino. Thus, reconstructing these final states is difficult. Furthermore, the $\tau \to \ell^-\nu\overline{\nu}$ decay is used frequently as it involves detecting the same lepton as in the denominator of Eq. (3.18), thus cancelling some systematic effects. Measuring $R_{D^*}$ alone

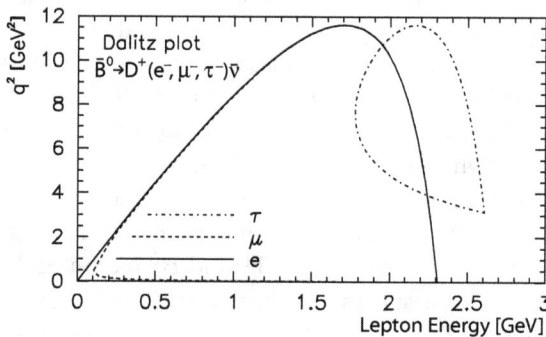

Fig. 3.21: The phase space in the $q^2$ versus lepton energy pseudo-Dalitz plane for $\overline{B}^0 \to D^+(e^-, \mu^-, \tau^-)\overline{\nu}$ decays plotted in the $\overline{B}^0$ rest frame. (Note, for $\tau^-$ the increased average energy compared to the lighter leptons is due to the much larger $\tau^-$ mass.) Adopted from Ref. [238].

is simpler than $R_D$ because the latter requires simultaneously determining the feed-down of $D^*$ mesons into $D$ mesons.

Both the BaBar [226, 227] and Belle [239] experiments have used hadronic $B$ tags to measure the 8 rates need to determine $R_D$ and $R_{D^*}$ implied by Eq. (3.18), specifically $D^0\tau^-\bar{\nu}$, $D^+\tau^-\bar{\nu}$, $D^{*0}\tau^-\bar{\nu}$, $D^{*+}\tau^-\bar{\nu}$, $D^0\ell^-\bar{\nu}$, $D^+\ell^-\bar{\nu}$, $D^{*0}\ell^-\bar{\nu}$, and $D^{*+}\ell^-\bar{\nu}$. They both use $\tau^- \to e^-/\mu^-\nu\bar{\nu}$ decays.[5] While the approaches of the two experiments are similar, the methods differ in important details and for a full understanding reading the published papers is necessary. Here we first discuss the general approach. The main discrimination variable is $\mathrm{MM}^2$ as defined in Eq. (3.16). The full reconstruction of the $B$ meson tag provides $\vec{p}_B = -\vec{p}_{\mathrm{tag}}$. One major problem is that the $D^{(*)}\tau^-\bar{\nu}$ signal does not peak. Quoting from Ref. [239], "the high $\mathrm{MM}^2$ region (above 0.85 GeV$^2$), where the $\tau^-$ signal is concentrated, exhibits little discrimination power between the $\tau^-$ signal and the other backgrounds, in particular, the $D^{**}$ background."

Backgrounds arise from several sources. One is $D^{**}$ mesons resulting from the allowed decays $\bar{B} \to D^{**}\ell^-\bar{\nu}$ and $\bar{B} \to D^{**}\tau^-\bar{\nu}$. The latter is suppressed due to the relatively large $\tau^-$ mass. The former is a particularly pernicious background as the presence of an additional missing pion or two from the $D^{**}$ decay, in conjunction with the neutrino, creates a final state that is quite similar to the sought after $D^{(*)}\tau^-\bar{\nu}$, $\tau^- \to \mu^-\nu\bar{\nu}$. signal. To study this background BaBar simulates the specific $D^{**}$ states $D_2^*(2460)$, $D_0^*(2300)$, $D_1(2420)$, $D_J(2580)$, and the radial excitations $D(2S)$ and $D^*(2S)$. As branching fractions for semileptonic decays with these resonances have not been measured, a theoretical model [241] is used for the decay rates and form-factors. Belle leaves out the radial excitations. Other allowed $D^{**}$ non-resonant systems, of $D^{(*)}\pi\pi$ are tested in the final fit by Belle. To help "control" these backgrounds separate samples of $D^*\pi^0\ell^-\bar{\nu}$ are also selected. Background from continuum $e^+e^- \to q\bar{q}$ interactions is also present. It is suppressed, but not eliminated by event shape considerations.[6] Note, there is not enough continuum data to do a direct subtraction without substantially increasing the statistical uncertainty, so simulation matched to the available data is used to model this component. Cross-feeds between

---

[5]Earlier measurements of semileptonic decays with $\tau$ leptons were made by both collaborations [240].

[6]The $e^+e^- \to q\bar{q}$ events appear as two collimated cones of particles, sometimes called jets, while the $\Upsilon(4S)$ decays are more spherical in shape.

different channels can change the charges of the signal decays and are treated as backgrounds.

Another large source of background are decays with two charmed hadrons in the final state where one of them decays into a lepton. These decays also possess final states with a $D$ or $D^*$ and a light lepton wiht the correct sign of charge. They are also modeled as part of the background fit.

Means of discrimination between signal and backgrounds are needed. For the situation at hand with one $B$ fully reconstructed and the other decaying into one of the 8 channels listed above, there should be no additional charged or neutral particles. Both the BaBar and Belle detectors are equipped with Thallium doped CsI crystal calorimeters based on the one developed for the CLEO II detector [17]. The calorimeters are extraordinarily efficient at measuring any extra energy ($E_{\text{extra}}$) not from the sought after decays. Unfortunately there are electromagnetic processes, such as $e^+e^- \to e^+e^-\gamma$, that deposit energy, as well as energetic fragments from interactions of the final state signal particles in the calorimeter, but a typical requirement that there be more than $50\,\text{MeV}$ of energy in each cluster eliminates most of them. For example, a previous measurement of $\mathcal{B}(D_s^+ \to \tau^+\nu)$ in $e^+e^- \to D_s^*D_s$ collisions, where one $D_s$ was fully reconstructed into hadronic final states, effectively used $E_{\text{extra}}$ primarily to discriminate signal from background [242].

The BaBar and Belle analyses differ substantially. We discuss the BaBar analysis first. They use as additional signal variables the momentum of the detected lepton in the $B$ rest frame ($|p_\ell^*|$) and $q^2$. These are correlated with the $\text{MM}^2$, but do provide additional information. For $\tau^-$ final states, $q^2$ must be larger than $m_\tau^2 = 3.1\,\text{GeV}^2$, so a cut of $>4\,\text{GeV}^2$ is imposed, which eliminates only a small part of the phase space (see Fig. 3.21). Then a neural network (BDT) is trained using these variables plus several others, most importantly the energy in the electromagnetic calorimeter that is not matched to any of the $B_{\text{tag}}$ or signal particles decays ($E_{\text{extra}}$) and $D$ and $D^*$ selection variables. After selecting on the BDT a fit to the data is performed with a total of 56 decay probability density functions (PDF's) describing signal and background components. Explicit references are not given to backgrounds from $\overline{B} \to D^{(*)}D_s^{-(*)}(X)$, $D_s^- \to \mu^-\nu$ or $D_s^- \to \tau^-\nu$, $\tau^- \to \mu^-/e^-\nu\overline{\nu}$, and $\overline{B} \to D^{(*)}D^{(*)}(X)$, $D \to Y\ell^-\overline{\nu}$ decays. These events are effective at faking $D^{(*)}\tau\nu$ as they contain missing particles. In the fits to the mass spectrum other background components could be taking care of this omission. The results are shown in Fig. 3.22.

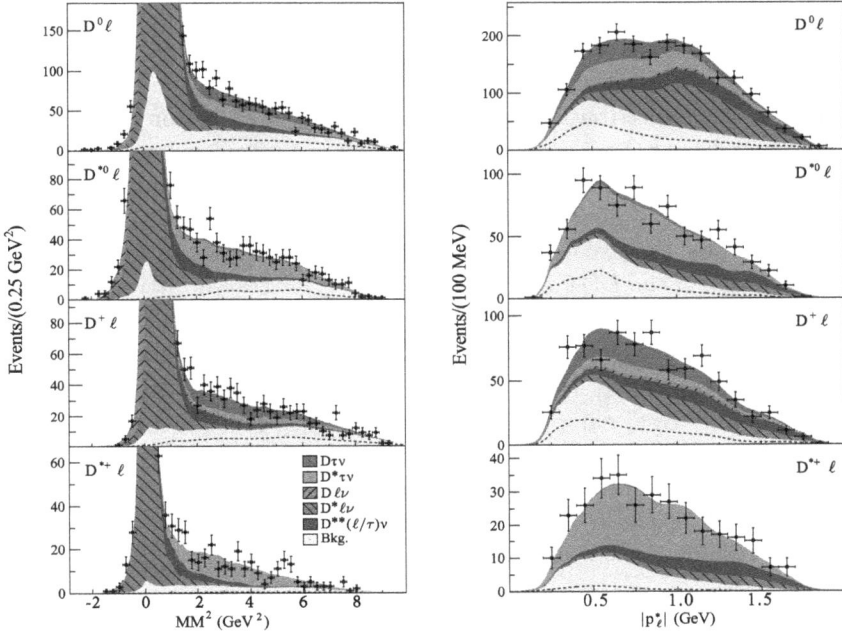

Fig. 3.22: (left) The resulting projection of the fit to the $MM^2$ distribution from BaBar [227]. The data points are given by black crosses. The resulting components are given by the color code on the bottom left. The peaks near zero are from $\overline{B} \to (D, D^*)\ell^-\overline{\nu}$. The lowest component (yellow area, labeled 'Bkg') indicates backgrounds from continuum that appear under the dashed line and cross-feeds that appear over the line and show peaking near $MM^2$ of zero. (right) Fit projection of $|p^*_\ell|$ for $MM^2 > 1\,GeV^2$.

Signals exist where they should, namely for the $D^*\tau^-\overline{\nu}$ final states there is little $D\tau^-\overline{\nu}$ (red), and the latter distributions contain both $D$ and $D^*$ due to the escape of pions or photons from $D^*$ decays. The $D^{**}$ background (brown/darkest-grey) is hard to distinguish from genuine $D^{(*)}\tau^-\overline{\nu}$ contributions by eye. BaBar [226, 227] finds

$$R_{D^*} = 0.332 \pm 0.024 \pm 0.018, \quad R_D = 0.440 \pm 0.058 \pm 0.042, \quad (3.19)$$

where the first uncertainty is statistical and the second systematic. The correlation coefficient is $-0.27$. The systematic error is a bit smaller than the statistical; the main contributions for $R_D$, $R_{D^*}$ are the $D^{**}$ contributions (5%, 2%) and simulation statistics (4.4%, 2%). Combining the statistical and systematic uncertainties in quadrature, and accounting for the correlation, results in a discrepancy with the SM prediction from the HFLAV average of $\sim 3.4\sigma$.

Next we will describe the Belle analysis [239]. They conceptually split the $MM^2$ distribution into a low mass-squared region $<0.85$ GeV$^2$ dominated by $D^{(*)}\ell^-\bar{\nu}$ decays and a high mass-squared region, which contains the $D^{(*)}\tau^-\bar{\nu}$ decays. They apply a BDT that optimizes their sensitivity using similar variables as BaBar especially $E_{extra}$. To train the BDT they use a combination of signal simulation and backgrounds. They transform the BDT output variable $o_{NB}$ for ease in fitting as $o'_{NB} \equiv \log[(o_{NB} - o_{NB}^{min})/(o_{NB}^{max} - o_{NB})]$. Then they fit simultaneously the entire $MM^2$ region and $o'_{NB}$ along with several other background shapes including cross-feeds, fake $D^{(*)}$, $\overline{B} \to D^{(*)}D_s^{-(*)}$ decays resulting in an $\ell^-$, and an additional "rest" category to capture any unspecific backgrounds they have not identified, and whose size and shape are fixed in the fit from the simulation. The resulting fit projections for the $D\ell^-$ modes are shown in Fig. 3.23 and for the $D^*\ell^-$ modes in Fig. 3.24.

The results of the overall fit are

$$R_{D^*} = 0.293 \pm 0.038 \pm 0.015, \quad R_D = 0.375 \pm 0.064 \pm 0.026. \qquad (3.20)$$

The correlation coefficient is $-0.56$. The systematic uncertainty is smaller than the statistical; two of the systematic error contributions for $(R_D, R_{D^*})$ are the $D^{**}$ (4.4%, 3.4%) and the $o'_{NB}$ shape (3.2%, 0.8%). Another relatively large error of 2.5% for both modes arises from the relative efficiency uncertainty. Combining the statistical and systematic uncertainties in quadrature, and accounting for the correlation, results in a discrepancy with the SM prediction from the HFLAV average of $\sim 1.8\sigma$, and differs from BaBar by $\sim 1.4\sigma$. Clearly these results are consistent with the SM.

The use of fully reconstructed hadronic tags allowed Belle to do another measurement of $R_{D^*}$ where the $\tau^-$ lepton decays into $\pi^-\bar{\nu}$ or $\rho^-\bar{\nu}$, $\rho^- \to \pi^-\pi^0$ [243]. Leptons in semileptonic decays are produced polarized in the SM because the weak interaction is left handed. This could be modified by NP effects. While there is no known method of measuring the lepton polarization in $\overline{B} \to D\tau^-\bar{\nu}$ when $\tau^- \to \ell^-\nu\bar{\nu}$, the $\tau^-$ decay angular distribution into two-body final states allows such a measurement. There are several predictions of NP effects [244–253].

We define the semileptonic decays with a $\tau^-$ polarized in the $+1/2$ state as $\Gamma^+(D^{(*)})$ and, similarly for the $-1/2$ state as $\Gamma^-(D^{(*)})$, where the $D$ or

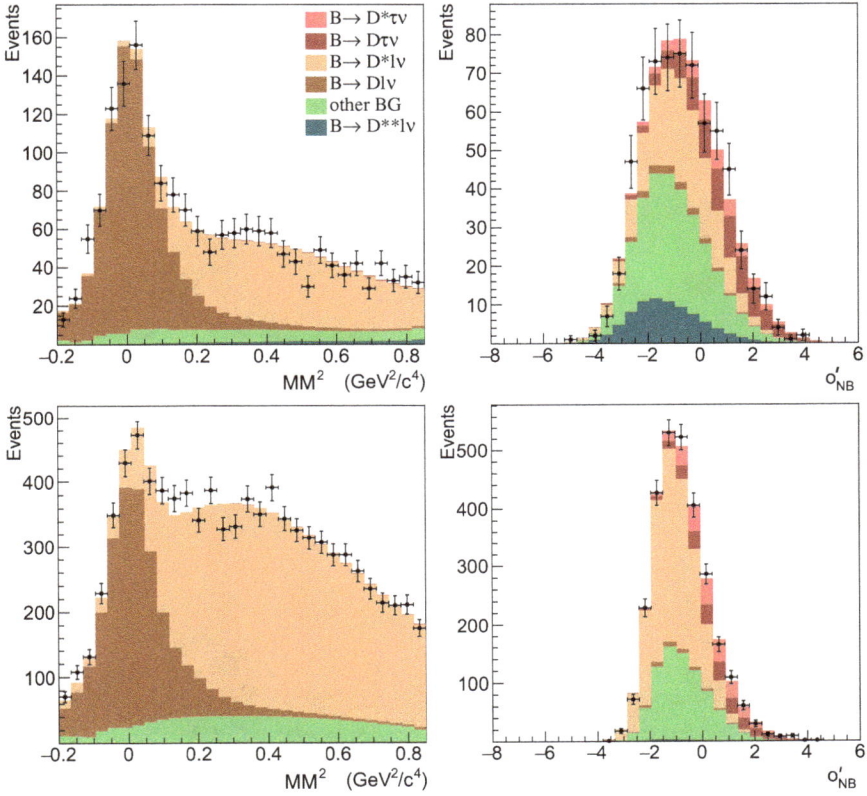

Fig. 3.23: Fit projections and data points with statistical uncertainties in the $D^+\ell^-$ (top) and $D^0\ell^-$ (bottom) data samples. (left) MM$^2$ distribution for MM$^2$ < 0.85 GeV$^2$; (right) $o'_{\rm NB}$ distribution for MM$^2$ > 0.85 GeV$^2$. From Belle data [239].

$D^*$ refer to the final state charm hadron in the final state. The definition of the $\tau^-$ lepton polarization is

$$P_\tau(D^{(*)}) = \frac{\Gamma^+(D^{(*)}) - \Gamma^-(D^{(*)})}{\Gamma^+(D^{(*)}) + \Gamma^-(D^{(*)})}. \tag{3.21}$$

In the SM $P_\tau(D) = 0.325 \pm 0.009$ [245] and $P_\tau(D^*) = -0.497 \pm 0.013$ [249].

$P_\tau(D^{(*)})$ can be measured by determining the $\overline{B} \to D^{(*)}\tau^-\overline{\nu}$ decay rate as function of the angle of the hadron from the $\tau^-$ decay with respect to the direction opposite the momentum of the $\tau^-\overline{\nu}$ system in the rest frame

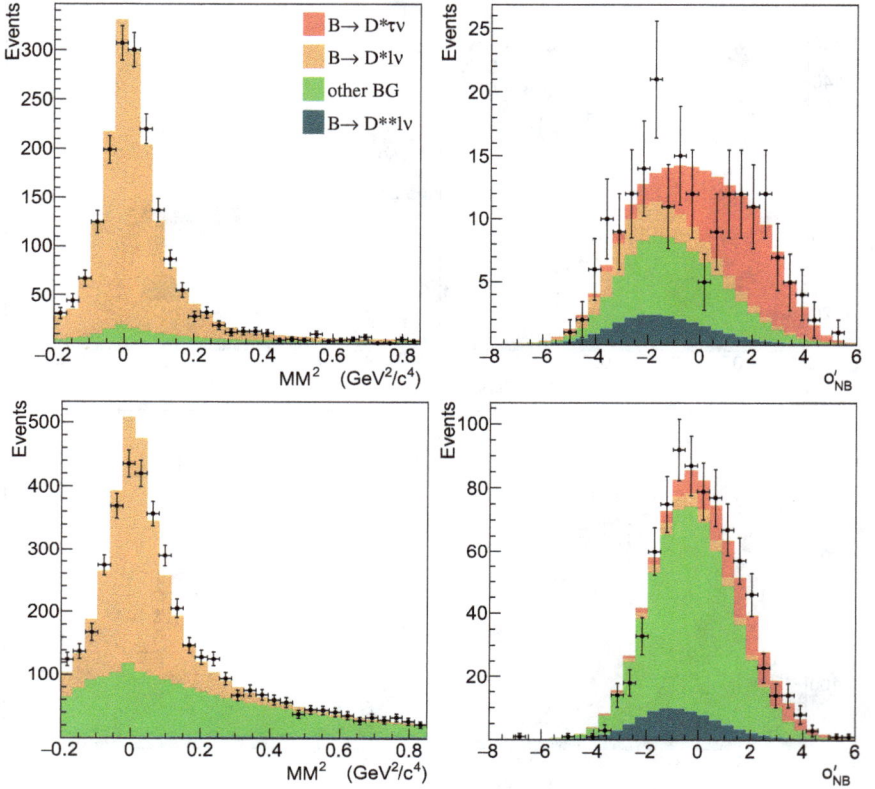

Fig. 3.24: Fit projections and data points with statistical uncertainties in the $D^{*+}\ell^-$ (top) and $D^{*0}\ell^-$ (bottom) data samples [239]. (left) MM$^2$ distribution for MM$^2$ < 0.85 GeV$^2$; (right) $o'_{\text{NB}}$ distribution for MM$^2$ > 0.85 GeV$^2$.

of $\tau^-$ ($\theta_{\text{hel}}$). The shape of the experimental distribution is characterized as

$$\frac{1}{\Gamma(D^{(*)})} \frac{d\Gamma(D^{(*)})}{d\cos\theta_{\text{hel}}} = \frac{1}{2}\left[1 + \alpha P_\tau(D^{(*)})\cos\theta_{\text{hel}}\right]. \tag{3.22}$$

Different decay modes have different values of $\alpha$. For $\tau^- \to \pi^-\nu$ $\alpha = 1$ and for $\tau^- \to \rho^-\nu$ $\alpha = 0.45$ [254]. It is possible to determine this angle because the signal $B$ meson four-momentum is known, as is the motion of the $\Upsilon(4S)$ center-of-mass so a Lorentz transformation is applied to bring the kinematic variables into the $\tau^-\bar{\nu}$ rest frame. This is not quite the $\tau^-$ frame but the difference is small enough so that at the current level of accuracy this is sufficient.

Here Belle does only the $D^*\tau^-\overline{\nu}$ final state, which is a large simplification. The analysis is split into two parts. The yields of the normalization modes are found by fitting to the $MM^2$ distributions very similar to the one described in the previous analysis. Since a main objective of this analysis is to measure $P_\tau(D^*)$, the data are analyzed in two bins of $\cos(\theta_{\text{hel}})$, one for $-1.0 < \cos(\theta_{\text{hel}}) < 0$, and the other for $0 < \cos(\theta_{\text{hel}}) < 0.8$. The upper limit is imposed due to large backgrounds from the normalization mode above that value. For the $\tau^-$ modes $q^2$ is again required to be above $4\,\text{GeV}^2$ and $E_{\text{extra}}$ below $1.5\,\text{GeV}$. Most of the same backgrounds are present as in the $R_{D^*}$ analysis presented above with the addition of other $\tau^-$ decays contributing and cross-feeds between the $\pi^-\overline{\nu}$ and $\rho^-\overline{\nu}$ modes. A fit to the data is performed using the signal variables $MM^2$, $q^2$, and $|\overrightarrow{p}^{\,*}_{\pi^-/\rho^-}|$, where the latter is the momentum of the hadron from the $\tau^-$ decay computed in the $\Upsilon(4S)$ rest frame. Backgrounds are represented by PDF's similar to the ones used above. The projections of the fits on these variables is shown in Fig. 3.25

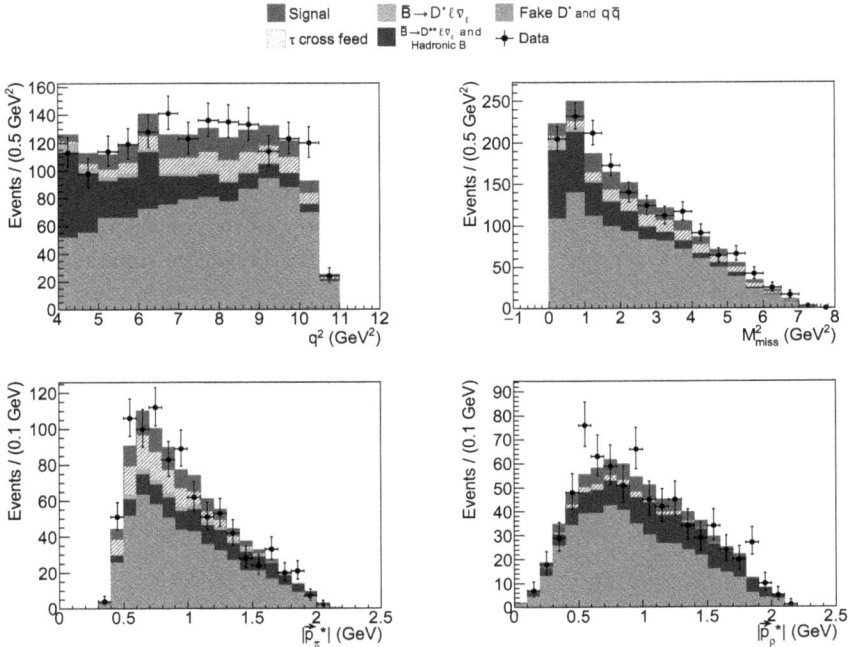

Fig. 3.25: Projections of the fit results on the distributions of $q^2$ (top-left), $M^2_{\text{miss}}$ (top-right), $|\vec{p}^{\,*}_\pi|$ (bottom-left) and $|\vec{p}^{\,*}_\rho|$ (bottom-right) from Belle [243]. These distributions contain the sums of both helicity samples.

The largest component of the fits is the fake $D^*$ and continuum backgrounds. Then there are similar event numbers in $D^{**}$ and hadronic $B$ backgrounds, the $\tau^-$ cross-feeds, and the $\tau^-$ signals. The results are

$$R_{D^*} = 0.270 \pm 0.035^{+0.028}_{-0.025} \quad P_\tau(D^*) = -0.38 \pm 0.51^{+0.21}_{-0.16}, \qquad (3.23)$$

where, as usual the uncertainties are statistical and systematic. The numbers are correlated with the correlation coefficient being $\sim$0.3. The largest systematic uncertainties for both measurements arise from the hadronic $B$ composition of the data including the $B$ tags and the statistics for the PDF shape. The $R_{D^*}$ measurement is well within $1\sigma$ of the SM prediction and the $P_\tau(D^*)$ has a large enough uncertainty that it is consistent with the entire range of physical possibilities. It does indicate, however, that this measurement can be made more precise with more data and simulation.

Finally, we note that the fraction of $D^{*+}$ longitudinal polarization in $\overline{B}^0 \to D^{*+}\tau^-\overline{\nu}$ has also been measured as $0.60 \pm 0.08 \pm 0.04$ by Belle [255] and is larger than and consistent with SM predictions [256, 257].

### 3.4.3 Measurements of $R_{D^{(*)}}$ using semileptonic decay tags at B factories

Pursuing $R_{D^{(*)}}$ further, Belle did another analysis where they use semileptonic $\overline{B} \to D^{(*)}\ell^-\overline{\nu}$ candidates for their tagged sample [258]. This has the advantage of a higher tagging efficiency than in hadronic decays, with the sacrifice of a good $\mathrm{MM}^2$ resolution. It is necessary to reduce the continuum background, which is done using event shape variables. The signal samples are the same as discussed above. In order to select the tags they define a new variable which is an alternative to $\mathrm{MM}^2$, the cosine of the angle between the detected $D^{(*)}\ell^-$ particles and the tagged $B$ in the $\Upsilon(4S)$ rest frame, $\cos\theta_{B,D^{(*)}\ell}$. At first glance this looks nonsensical since the direction of the tagged $B$ is not known, however assuming there is only a missing neutrino we have $E_{\mathrm{beam}} = E_B = E_{D^{(*)}\ell} + E_\nu$, and $\vec{p}_B = \vec{p}_{D^{(*)}\ell} + \vec{p}_\nu$, where $D^{(*)}\ell$ means the measured sum of the $D^{(*)}$ and $\ell^-$. Next we compute $m_B^2 = E_{\mathrm{beam}}^2 - \vec{p}_B^2$, and solving the resulting equation we arrive at

$$\cos\theta_{B,D^{(*)}\ell} = \frac{2E_{\mathrm{beam}}E_{D^{(*)}\ell} - m_B^2 - m_{D^{(*)}\ell}^2}{2|\vec{p}_B||\vec{p}_{D^{(*)}\ell}|}, \qquad (3.24)$$

where all the quantities on the right-side of the equation are either known ($E_{\mathrm{beam}}, |\vec{p}_B|$) or measured for each candidate. (Note, $|\vec{p}_B|^2 = E_{\mathrm{beam}}^2 - m_B^2$).

Correctly identified tag candidates should be in the range $|\cos\theta_{B,D^{(*)}\ell}| < 1$, however due to resolution Belle keeps as signal events those with $-2 < \cos\theta_{B,D^{(*)}\ell} < 1$; the backgrounds peak at even more negative values. Then signal $D^{(*)}\tau^-/\ell^-\overline{\nu}$ candidates are searched for. To prevent confusion between tag and signal candidates only opposite flavor is allowed. This removes some mixed $B^0$ events, but significantly reduces backgrounds. In each event, $E_{\text{extra}}$ in the calorimeter is required to $<1.2\,\text{GeV}^2$, where low energy clusters, $\sim100\,\text{MeV}$ or less, are not counted in the energy sum as they are dominated by beam related backgrounds. A BDT is formed using as input 1) $\cos\theta_{B,D^{(*)}\ell}$, 2) the signal side $MM^2$ from Eq. (3.17), and 3) the sum of all the visible energy in the event. The BDT output for each event is used as a fitting variable as is $E_{\text{extra}}$. Each of the $D^{(*)}\tau^-/\ell^-\overline{\nu}$ samples has its own BDT.

The results of this analysis are

$$R_{D^*} = 0.283 \pm 0.018 \pm 0.014, \quad R_D = 0.307 \pm 0.037 \pm 0.016, \quad (3.25)$$

where the two results have statistical and systematic correlations of $-0.53$ and $-0.52$. The largest components of the systematic uncertainties are $(R_{D^*}, R_D)$ MC statistics (2.25%, 4.39%) and efficiency factors (4.12%, 2.25%). These results are compatible with the SM at the level of $0.8\sigma$, and cannot be directly averaged with previous Belle hadronic tag results because they use the same signal sample.

### 3.4.4 *Measurements of $R_{D^*}$ and $R_{J/\psi}$ at LHCb*

At first glance, the LHC appears to be a very hostile environment in which to make a measurement with multiple missing particles in a heavy flavor decay. First of all there are on average $\sim50$ tracks from each $pp$ interaction and there can be more than one in each beam crossing. Secondly, finding final states with electrons is much harder than using muons due to the electron bremsstrahlung, so LHCb restricts their measurements to be based only on muons. The only reason it is possible to attempt these measurements is due to the excellent ability of LHCb to distinguish decays of "long-lived" particles from the primary $pp$ interaction position, called the PV, and each other. This is easier explained by referring to sketches. Figure 3.26 shows the topologies of the decays (a) $\overline{B}^0 \to D^{*+}\mu^-\overline{\nu}$, (b) $\overline{B}^0 \to D^{*+}\tau^-\overline{\nu}$, $\tau^- \to \mu^-\nu\overline{\nu}$, and (c) $\overline{B}^0 \to D^{*+}\tau^-\overline{\nu}$, $\tau^- \to \pi^-\pi^-\pi^+\nu$. Candidate events with well defined decay points provide a determination of the $\overline{B}^0$ direction. In (a) for example, the $D^0$, which decays into $K^-\pi^+$

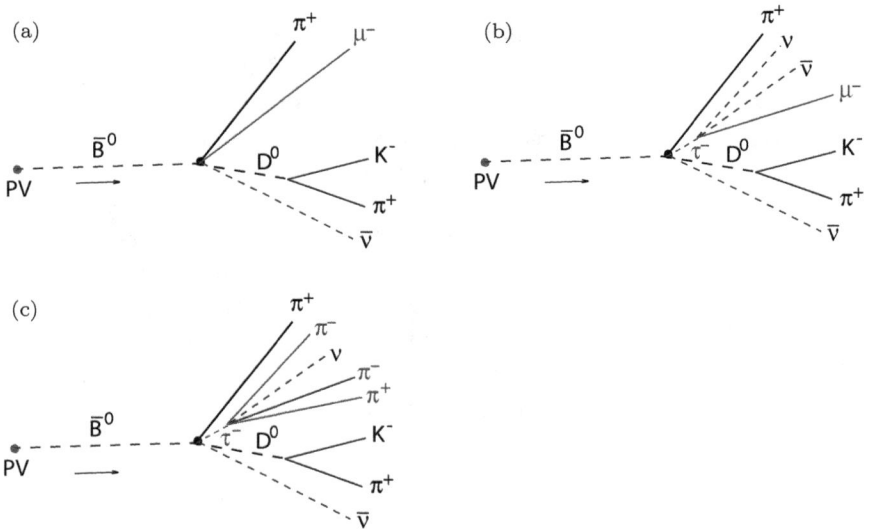

Fig. 3.26: Sketches of the decay topologies for (a) $\bar{B}^0 \to D^{*+}\mu^-\bar{\nu}$, (b) $\bar{B}^0 \to D^{*+}\tau^-\bar{\nu}$, $\tau^- \to \mu^-\nu\bar{\nu}$ and (c) $\bar{B}^0 \to D^{*+}\tau^-\bar{\nu}$, $\tau^- \to \pi^-\pi^-\pi^+\nu$. The dashed lines indicate particles that do not make visible hits in the vertex detector.

and the $\mu^-$ come from the same decay point. So does the $\pi^+$ from the $D^{*+} \to D^0$ decay, but it is less precise at determining the decay vertex point, because it multiple scatters more due to its lower momentum. The PV is well defined by the large number of tracks and so we have a measurement of the $\bar{B}^0$ direction. Its precision depends on several factors including the distance of the $\bar{B}^0$ decay from the PV and decay configuration of the other particles. The $\bar{B}^0$ usually travels far before decaying, for example, since the lifetime is 1.52 ps, a typical 50 GeV $\bar{B}^0$ meson travels ~0.5 cm, and the spatial resolution of the tracking detector is about 10 microns in directions perpendicular to the beam-line. LHCb has done several important studies with semileptonic $\Lambda_b^0$ [72, 94] and $B_s^0$ decays [259, 260].

The decay with $\tau^- \to \mu^-\nu\bar{\nu}$, shown in (b), has the $\mu^-$ appear after the $\bar{B}^0$ decay point, but not by a lot, as the $\tau^-$ lifetime is only 0.29 ps, and the $\tau^-$ has much smaller momentum than the $\bar{B}^0$. In fact, it is assumed that that the $D^{*+}$ and the $\mu^-$ define the $\bar{B}^0$ decay point. This causes the vertex position to be somewhat smeared, but it is still workable. This mode is used because it relatively easy to identify the muon. In case (c) the $\tau^-$ has a discernible decay vertex because it decays into three pions that are constrained to come from a single decay point, and the $\bar{B}^0$ vertex is determined by the $D^{*+}$ decay point.

We first discuss the ratio using the $\tau^- \rightarrow \mu^- \nu \bar{\nu}$ decay [261]. A neural network is used to ensure that there are no other tracks that are consistent with coming from the $D^{*+}\mu^-$ vertex. This requirement eliminates some background. Signal event selection relies on separating signal from remaining background using three kinematic variables defined in the $\bar{B}^0$ rest frame: $\text{MM}^2$ [Eq. (3.15)], the energy of the $\mu^-$ ($E_\mu^*$), and $q^2 = (p_B^\mu - p_{D*}^\mu)^2$, in four-vector notation. $\text{MM}^2$ is the best variable to measure the rate of the normalization mode $\bar{B}^0 \rightarrow D^{*+}\mu^-\bar{\nu}$. Since in the $\tau^-$ mode there are two extra neutrinos, $E_\mu^*$ will have smaller values than in the $\mu^-$ mode. Finally, the $q^2$ must be larger than $m_\tau^2$.

Obviously, the $\bar{B}^0$ momentum cannot be measured directly, but it can be approximated. The LHCb spectrometer is centered on the beamline which is almost parallel to the $+z$ direction (see section 1.3). The z-component of the $\bar{B}^0$ momentum can be approximated as

$$[p_B]_z = \frac{m_B}{m_{\text{reco}}}[p_{\text{reco}}]_z, \tag{3.26}$$

where $m_B$ is the known $\bar{B}^0$ mass, $m_{\text{reco}}$ is the reconstructed invariant mass of the $D^{*+}$ and the $\mu^-$, and $[p_{\text{reco}}]_z$ is the $z$ component of their combined momentum. Since the angle that the $\bar{B}^0$ makes with the $z$ direction is known, the lab momentum of the $\bar{B}^0$ is found and the various particle four-momenta can be transformed to the $\bar{B}^0$ rest frame in order that the discrimination variables can be computed.

Templates are constructed from simulation for the $\bar{B}^0 \rightarrow D^{*+}\tau^-\bar{\nu}$ signal as well as the $\bar{B}^0 \rightarrow D^{*+}\mu^-\bar{\nu}$ normalization mode. While the $\bar{B}^0 \rightarrow D^{*+}$ transition form factors are well known for the normalization mode, there are only theoretical calculations for the signal mode based on sum rules. So the fit template uses these values and includes a constraint based on the estimated model uncertainty.

There are several parameterized backgrounds. Final states that contain $D^{**}$ that are known to decay into a $D^{*+}$ and a $\pi$, the $D_1(2420)$, $D_2^*(2460)$, $D_1'(2430)$ are included, as well as $\bar{B}_s^0 \rightarrow [D_{s1}'(2536)^+$ and $D_{s2}(2573)^+]\mu^-\bar{\nu}$ decays [262]. The form-factors for these decays are taken from models. A $\bar{B}^0 \rightarrow D^{**}\mu^-\bar{\nu}$, $D^{**} \rightarrow D^{*+}\pi\pi$ shape is also included. Estimates of the branching fractions for these modes allow a fixed value with an uncertainty to be assigned to these states for use in determining the size of the normalization mode. For the signal mode since branching fractions are unknown, as well as the form-factors, rates proportional to the normalization mode are assumed. In the case of the $\overline{B}_s^0$ decay backgrounds the non-resonant $\overline{B}_s^0 \to DKX\mu^-\overline{\nu}$ decays are not directly accounted for, which should approximately double the $\overline{B}_s^0$ background estimate [263]. $\Lambda_b^0 \to D^{*+}nX\mu^-\overline{\nu}$ decays are expected to be small from an upper limit on the isospin related channel. Compensation of these omissions and the lack of knowledge of $D^{**}$ production in muonic and tauonic semileptonic decays is taken care of by allowing the yields of $\overline{B} \to D^{**}\mu^-\overline{\nu}$, $D^{**} \to D^{*+}\pi\pi$ and $\overline{B}_s^0 \to D_s^{**+}\mu^-\overline{\nu}$, $D_s^{**+} \to D^{*+}K_S^0$ to float in the final fit to the data, and use of a $D^{*+}\mu^-\pi^+\pi^-\overline{\nu}$ control sample. Another control sample $D^{*+}\mu^+$ is used to measured the background under the $D^{*+}$ peak. A significant background is from $b \to D^{*+}DX$ events, where $D \to Y\mu\overline{\nu}$. These shapes are also modeled in simulation that match with control samples defined in the data. Other backgrounds come from either $D_s^- \to \tau^-\nu$, $\tau^- \to \mu^-\nu\overline{\nu}$ decays or $D_s^- \to \mu^-\overline{\nu}$ decays.

Many quantities are extracted from the overall fit besides $R_{D^*}$. They include The $D^{**}$ yields and their form-factors; $\overline{B}_s^0 \to D^{*+}K_s\mu^-\overline{\nu}$ yields, and background yields. The results of the fit are shown in Fig. 3.27.

The $D^{*+}\tau^-\overline{\nu}$ (red) appears in the highest four $q^2$ bins as expected. There are large backgrounds there from $B$ decays with a $D^{*+}$ and an addition charmed hadron that decayed into a muon (green) as sell as a significant $D^{**}$ background. The largest component everywhere is the normalization decay (dark blue). Combinatorial background is also shown (yellow), with muon misidentification (orange) being quite small. The result is

$$R_{D^*} = 0.336 \pm 0.027 \pm 0.030. \qquad (3.27)$$

The largest systematic uncertainties are the simulation statistics (6.0%) and the template shape for muon misidentification (4.8%).

LHCb has also measured $R_{D^*}$ with $\tau^- \to \pi^+\pi^-\pi^-\overline{\nu}$ decays [264]. The decay chain is sketched in Fig. 3.26(c). The $\overline{B}^0$ vertex is defined by the intersection of the $\pi^+$ from the $D^{*+}$ decay with the related $D^0$ decay. The $\pi^+\pi^-\pi^-$ ($3\pi$) is detected by requiring the tracks to form a separate vertex downstream of the $\overline{B}^0$ decay point. Instead of directly measuring $\mathcal{B}\left(\overline{B}^0 \to D^{*+}\mu^-\overline{\nu}\right)$ they use the decay $\overline{B}^0 \to D^{*+}\pi^+\pi^-\pi^-$ which they separately measure and use the ratio of these branching fractions from the PDG [12]. This introduces some further systematic uncertainty due to the current precision on the branching fractions.

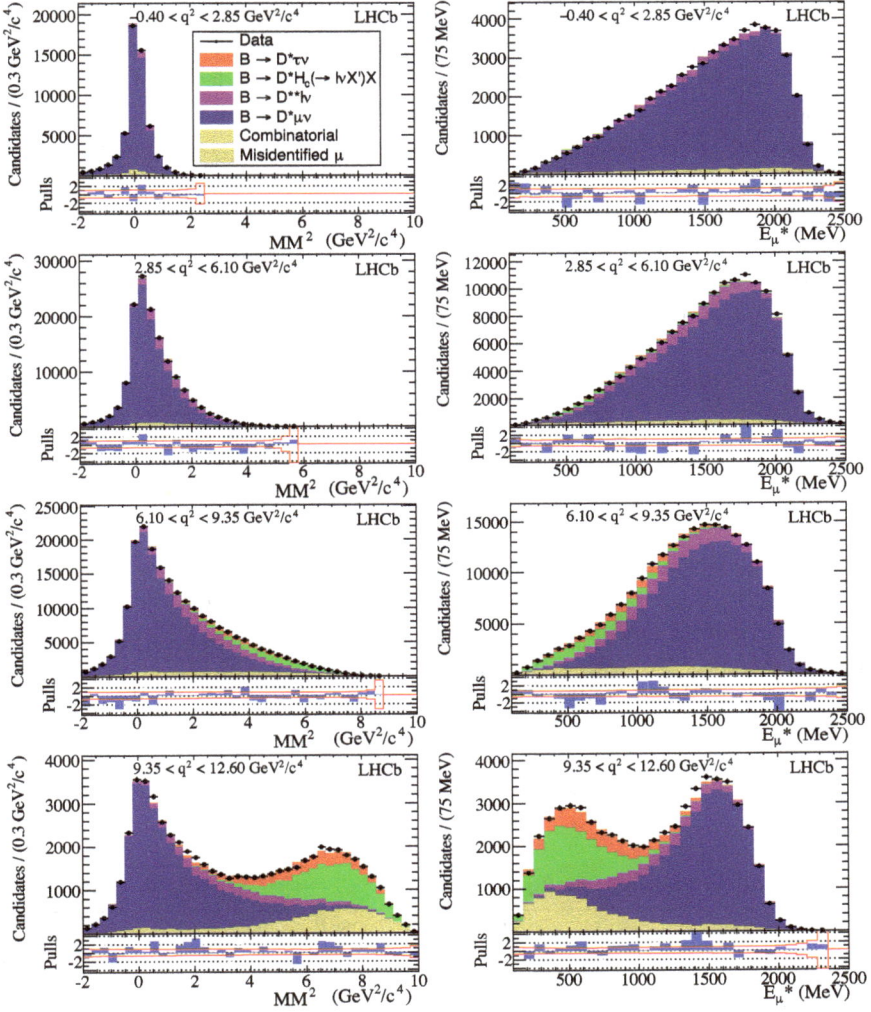

Fig. 3.27: Overall fit to the data shown as points with error bars to two variables $MM^2$ (left) and $E_\mu^*$ (right) in 4 different bins of $q^2$ [261]. The individual fit components are shown as colored bands. The legend is in the upper left sub-figure. The sub-plots labeled pulls show the deviation of the data from the fit.

Signal selection is implemented using a neural network (BDT) with requirements on goodness of the vertex fits, track transverse momenta, etc. To extract signal and background components, the data are fit in the three-dimensions of $q^2$, $(3\pi)$ decay time, which for signal should have an exponential distribution following the $\tau$ lifetime, and BDT output. In order to do this the first two variables need to be evaluated which is not straightforward as two neutrinos are missing in the signal decay. A significant portion of the $\tau^- \to \pi^- \pi^+ \pi^- \pi^0$ decays are accepted by the BDT and its efficiency is separately evaluated and included in the branching ratio calculation.

The measured directions of the $\bar{B}^0$ and the $\tau^-$ can be determined. The line between the PV and $\bar{B}^0$ vertex defines the $\bar{B}^0$ direction, while the line between the $3\pi$ vertex and the $\bar{B}^0$ vertex defines the direction of the $\tau^-$. This permits the calculation of $q^2$ and the $\tau^-$ lifetime with a four-fold ambiguity.

The $\tau^-$ momentum is given by

$$|\vec{p}_\tau| = \frac{(m_{3\pi}^2 + m_\tau^2)|\vec{p}_{3\pi}|\cos\theta_{\tau,3\pi} \pm E_{3\pi}\sqrt{(m_\tau^2 - m_{3\pi}^2)^2 - 4m_\tau^2|\vec{p}_{3\pi}|^2\sin^2\theta_{\tau,3\pi}}}{2(E_{3\pi}^2 - |\vec{p}_{3\pi}|^2\cos^2\theta_{\tau,3\pi})},$$

(3.28)

where $\theta_{\tau,3\pi}$ is the angle between the $3\pi$ system three-momentum and the $\tau$ line of flight; $m_{3\pi}$, $|\vec{p}_{3\pi}|$ and $E_{3\pi}$ are the mass, three-momentum and energy of the $3\pi$ system, and $m_\tau$ is the known $\tau$ mass. The two-fold ambiguity is removed simply by setting the square root term to zero.

Similarly the magnitude of the $\bar{B}^0$ momentum can be found. First defining

$$\vec{p}_Y = \vec{p}_{D^{*+}} + \vec{p}_\tau \quad E_Y = E_{D^{*+}} + E_\tau,$$

(3.29)

we have

$$|\vec{p}_{B^0}| = \frac{(m_Y^2 + m_{B^0}^2)|\vec{p}_Y|\cos\theta_{B^0,Y} \pm E_Y\sqrt{(m_{B^0}^2 - m_Y^2)^2 - 4m_{B^0}^2|\vec{p}_Y|^2\sin^2\theta_{B^0,Y}}}{2(E_Y^2 - |\vec{p}_Y|^2\cos^2\theta_{B^0,Y})}.$$

(3.30)

Here, the three-momentum and mass of the $D^{*-}\tau$ system are calculated using the previously estimated $\tau$ momentum. Again, the square root term is set to zero. Using this method, the rest frame variables $q^2 \equiv (p_{B^0} - p_{D^{*-}})^2 = (p_\tau + p_{\nu_\tau})^2$ and the $\tau$ decay time, $t_\tau$, are determined with sufficient accuracy to retain their discriminating power against double-charm backgrounds. The relative $q^2$ resolution is 18% full-width half-maximum.

The most important backgrounds here have some similarity with the previous LHCb analysis using the $\tau^- \to \mu^- \nu\bar{\nu}$ decay mode, namely

$b \rightarrow D^{*+}h_c(X)$, where $h_c$ indicates a charm hadron and $(X)$ the possible presence of additional pions or kaons. This background has been investigated and normalized to the specific channels $\bar{B}^0 \rightarrow D^{*+}D_s^-(X)$ by measuring a sample where $D_s^- \rightarrow \pi^-\pi^+\pi^-$. Other backgrounds from $h_c \equiv D^+$, $D^0$ are normalized in a similar manner. Many checks are done with special data samples, such as $D^{*+}3\pi$ where the $3\pi$ and the $D^{*+}$ form a vertex. The results of these studies are entered into the final fit as predicted event rates with Gaussian uncertainties and simulated shapes. For the $D^{**}\tau^-\bar{\nu}$ backgrounds they merely rely on theoretical calculations and fix the rate as $(11\pm4)\%$ of the signal. The magnitude of this background as well its assigned uncertainty are questionable.

The results of the fit are shown in Fig. 3.28. The fit appears quite good with the $D^{*+}\tau^-\nu$ signal components appear increasing at large $q^2$. The largest systematic uncertainties are due to simulation statistics (4.1%) and four background models each about 2.6%.

The result is

$$R_{D^*} = 0.291 \pm 0.019 \pm 0.026 \pm 0.013, \tag{3.31}$$

where the last uncertainty is due to $\mathcal{B}(\bar{B}^0 \rightarrow D^{*+}\pi^+\pi^+\pi^-)$. The result is $1.1\sigma$ above the SM prediction.

Another related result is the LHCb measurement [265] called $R_{J/\psi} \equiv \mathcal{B}(B_c^- \rightarrow J/\psi\tau^-\bar{\nu})/\mathcal{B}(B_c^- \rightarrow J/\psi\mu^-\bar{\nu})$. The ratio of $B_c^-$ production to all $b$'s is only $\approx 0.26\%$ [266], so statistics will be a limitation. On the other hand, the $J/\psi$ decays at the $B_c^-$ decay point providing a well determined decay vertex, and the backgrounds are very different than in other studies. They contain feed-downs from higher charmonium states to the $J/\psi$ and the modes $B_c^- \rightarrow J/\psi D_s^-(X)$, plus combinations of the two. The final result is

$$R_{J/\psi} = 0.71 \pm 0.17 \pm 0.18. \tag{3.32}$$

The result has a $3\sigma$ statistical significance and is $2\sigma$ above the SM prediction. Only $3\,\text{fb}^{-1}$ of 7 and $8\,\text{TeV}$ data was analyzed, so this should become more interesting when the factor of four remaining data taken at $13\,\text{TeV}$ is analyzed.

### 3.4.5 *Summary of $R_D$ and $R_{D^*}$*

Figure 3.29 shows the summary of $R_{D^{(*)}}$ results prepared by the HFLAV data averaging collaboration [74] in the Spring of 2019 is applicable to all results as of August of 2021, and possibly later.

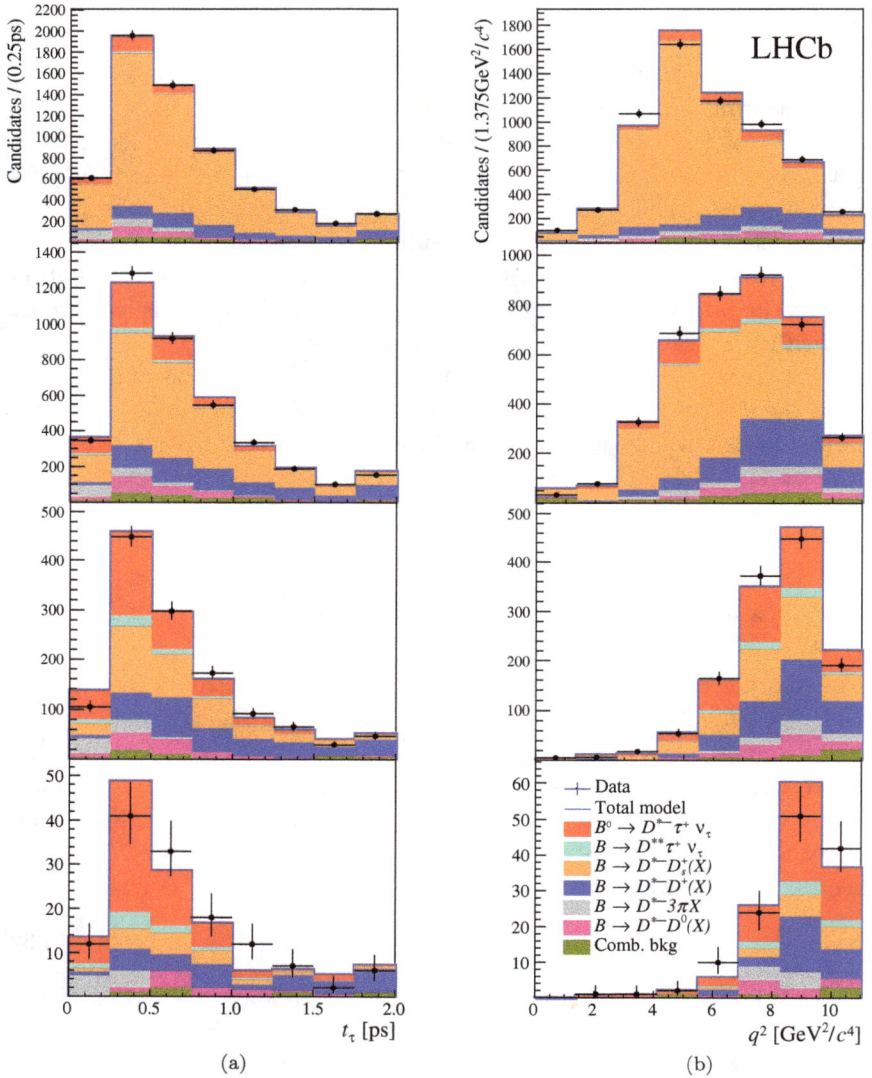

Fig. 3.28: (left) Reconstructed $\tau^-$ decay time distributions ($t_\tau$) and (right) $q^2$ distributions for all the fit components shown in different BDT intervals that increase (more signal like) from top to bottom. From Ref. [264].

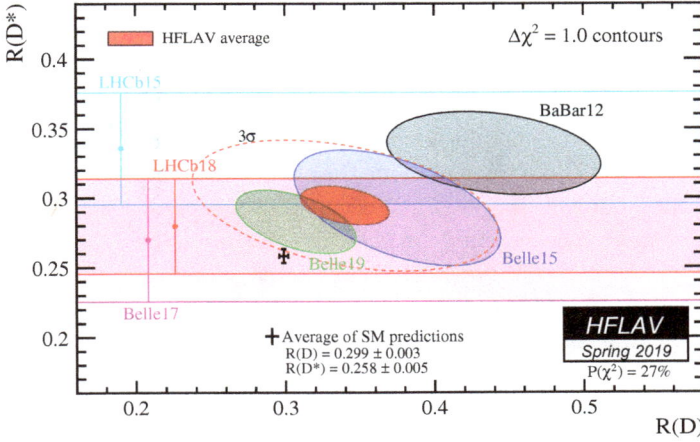

Fig. 3.29: Results of BaBar, Belle and LHCb $R_{D^{(*)}}$ measurements. Measurements of both $R_D$ and $R_{D^*}$ are shown as ellipses, while measurements of $R_{D^*}$ are shown has vertical lines near the left of the plot. The fitted average is shown as a red ellipse and the average of SM predictions as a cross, derived by HFLAV [74]. The P($\chi^2$)=27% refers to the consistency of the experimental measurements.

The HFLAV averages are

$$R_{D^*} = 0.295 \pm 0.010 \pm 0.010, \quad R_D = 0.340 \pm 0.026 \pm 0.014, \quad (3.33)$$

where the correlation between the results is $-0.38$. In the average HFLAV assumes a 100% correlation for systematic uncertainties associated with $D^{(*)}$ form factors, $D^{**}$ composition and form factor shapes, and the $\tau$ branching fractions. Other uncertainties are considered uncorrelated between the various experiments. The results show that $R_{D^*}$ exceeds the SM prediction by $2.5\sigma$ and $R_D$ by $1.4\sigma$. These results combined are $3.1\sigma$ above the SM predictions.

The experimental situation remains unclear. While all results are above the SM expectations the error bars on most are large enough so that each of the results are consistent within $2\sigma$ with the SM, the exception being the BaBar 2012 result. These are very difficult measurements both at $e^+e^-$ $B$-factories and at LHCb. The most serious concern is that the background sources have not been accurately estimated. We show in section 2.1.3 that the decay rate of inclusive semileptonic decays into electrons or muons exceeds the measured exclusive rates by about 10% implyin that there can be unknown sources of uncertainty. Specifically, only a few of the $D^{**}$ contributions to the muonic or electronic decays have been measured, but none in the tauonic modes. In both cases their form-factors need to be taken from theory. For another example of possible uncertainties, consider the $b \to D^{(*)}D(X)$ modes where the $D$ decays semileptonically. Some experiments only consider the two-body modes, while the measured branching fraction for the average of $\bar{B}^0$ and $B^+$ modes is $(8.3 \pm 0.8)\%$ [12], a larger number than what has been used, presumably due to three-body $b$ decays. Furthermore, in searches using $\tau^- \to \ell^- \nu \bar{\nu}$, it appears most experiments consider the effects of the $D_s^-$ only through $\mathcal{B}(D_s^- \to \tau^- \bar{\nu}, \ \tau^- \to \ell^- \nu \bar{\nu}) = 1.5\%$, while $\mathcal{B}(D_s^- \to \mu^- \nu) = 0.5\%$ adds a $1/3$ increase in this dangerous rate. In many cases fits to control samples are used to constrain these backgrounds. While the experiments may have covered these problems with sufficient systematic uncertainties, it is possible that one more backgrounds have not been included properly and they are nearly fully correlated among experiments. This would explain simply the fact that currently all experiments are on the high side of the SM predictions. On the other hand, these are important measurements and must be pursued with more data, more ancillary measurements, and possibly also more related channels.[7] Further theory advances also would be helpful. In this respect, new methods to reduce the form-factor ratio dependence have been proposed in Ref. [238, 268].

---

[7]A comprehensive review of these measurements is given in Ref. [267].

# Chapter 4

# Theoretical models addressing the current anomalies

## 4.1 Introduction

The experimental results illustrated in chapter 3 may represent a fundamental step forward in revealing the nature of physics beyond the Standard Model. Here we discuss some of the theoretical ideas which have been proposed in the last few years to interpret these anomalies and possibly link them to some of the open problems of the SM. Since many aspects are still unclear, and the field is in rapid evolution, we address the problem via a bottom-up perspective. We start employing generic effective-theory approaches based on few theoretical hypotheses, along the lines discussed in section 1.4.4, and only later on we add more hypotheses and discuss more explicit NP constructions.

More precisely, in section 4.2 and 4.3 we describe separately neutral-current and charged-current processes. Each set of anomalies is analysed employing a general Effective Field Theory (EFT) approach based only on the hypothesis that new physics is heavy. This approach allow us to combine different observables based on the same underlying partonic transition. In section 4.2 we show how the combination of data on $R_K$ and $R_{K^*}$, with those on $\mathcal{B}(B_s \to \mu^+\mu^-)$, leads to a strong evidence of new physics in $b \to s\ell\ell$ amplitudes. In a similar fashion, in section 4.3 we combine all available data on observables sensitive to $b \to c\tau\nu$ (and $b \to u\tau\nu$) amplitudes. In section 4.4 we go one step forward discussing how both set of anomalies can be described in a common EFT framework, addressing in a consistent way not only the anomalies, but also the tight bounds from a series of other low- and high-energy observables.

In section 4.5 we address the question of the possible mediators of these anomalies, focusing in particular on the case favored by the combined analysis, namely the $U_1$ leptoquark. Finally, we briefly discuss more ambitious ultraviolet (UV) complete models able to connect the two set of anomalies, not only among themselves, but also to some of the open problems of the SM.

## 4.2 The $b \to s\ell^+\ell^-$ anomalies

### 4.2.1 *The effective $b \to s\ell^+\ell^-$ Lagrangian*

Following the general discussion in section 1.4.2, in order to describe rare $B$ decays it is convenient to introduce an appropriate low-energy effective Lagrangian written in terms of light SM fields. This approach allow us to treat systematically the effects of all the heavy degrees of freedom present in the SM, such as the $W$ boson and the top quark, as well as the effects of generic heavy new physics.

The effective Lagrangian describing $B$ decays induced by a partonic transition of the type $b \to s\ell^+\ell^-$ can be decomposed in as

$$\mathcal{L}_{\text{eff}}^{b \to s} = \mathcal{L}_{\text{FCNC}}^{b \to s} + \mathcal{L}_{(4q)}^{b \to s}, \tag{4.1}$$

where

$$\mathcal{L}_{\text{FCNC}}^{b \to s} = \frac{4G_F}{\sqrt{2}} \left[ V_{tb}V_{ts}^* \sum_{i=7}^{10} C_i Q_i + \mathcal{O}(V_{ub}V_{us}^*) \right] + \text{h.c.} \tag{4.2}$$

$$\mathcal{L}_{(4q)}^{b \to s} = \frac{4G_F}{\sqrt{2}} \left[ V_{tb}V_{ts}^* \left( \sum_{i=1}^{2} C_i Q_i^c + \sum_{i=3}^{10} C_i^{(4q)} Q_i^{(4q)} \right) + \mathcal{O}(V_{ub}V_{us}^*) \right] + \text{h.c.} \tag{4.3}$$

and where $V_{ij}$ denote the elements of the Cabibbo-Kobayashi-Maskawa (CKM) matrix. The operators in $\mathcal{L}_{\text{FCNC}}^{b \to s}$ are bilinear operators in the quark fields, denoted in the following FCNC operators, while the operators in $\mathcal{L}_{(4q)}^{b \to s}$ are four-quark operators.

Within the SM, the set of relevant FCNC operators consists of four independent terms

$$Q_7 = \frac{e}{16\pi^2} m_b(\bar{s}_L \sigma^{\mu\nu} b_R) F_{\mu\nu}, \quad Q_8 = \frac{g_s}{16\pi^2} m_b(\bar{s}_L \sigma^{\mu\nu} T^a b_R) G_{\mu\nu}^a$$

$$Q_9^\ell = \frac{e^2}{16\pi^2} (\bar{s}_L \gamma_\mu b_L) \sum_\ell (\bar{\ell} \gamma^\mu \ell), \quad Q_{10}^\ell = \frac{e^2}{16\pi^2} (\bar{s}_L \gamma_\mu b_L) \sum_\ell (\bar{\ell} \gamma^\mu \gamma_5 \ell). \tag{4.4}$$

Among them, $Q_{7,9,10}$ are particularly relevant since they have, in general, a non-vanishing tree-level matrix element in $b \to s \ell^+ \ell^-$ transitions. Among the four-quark operators, those with the largest Wilson coefficients are

$$Q_1^c = (\bar{s}_L \gamma_\mu T^a c_L)(\bar{c}_L \gamma^\mu T^a b_L), \quad Q_2^c = (\bar{s}_L \gamma_\mu c_L)(\bar{c}_L \gamma^\mu b_L). \tag{4.5}$$

The complete list of SM operators, with the corresponding values of the Wilson coefficients, are reported in appendix C.

In extensions of the SM with new heavy degrees of freedom, we can expect possible modifications of the Wilson coefficients of the operators already present in the SM, but also the appearance of new effective operators. Still, we can maintain the general decomposition between FCNC operators and four-quark operators in Eq. (4.1). In the case of $\mathcal{L}_{\text{FCNC}}^{b \to s}$, we need to extend the operator basis including all the operators obtained by those in Eq. (4.4) replacing a left-handed field with a right-handed field (and vice versa). In principle, also operators involving scalar, pseudoscalar and tensor current could appear. Limiting the attention to terms which correspond to a local four-fermion interaction between two quarks and two lepton fields at the electroweak scale, which respects the gauge symmetry of the SM, the additional set of operators we must include is given by

$$Q_9'^\ell = \frac{e^2}{16\pi^2}(\bar{s}_R \gamma_\mu b_R) \sum_\ell (\bar{\ell} \gamma^\mu \ell), \quad Q_{10}'^\ell = \frac{e^2}{16\pi^2}(\bar{s}_R \gamma_\mu b_R) \sum_\ell (\bar{\ell} \gamma^\mu \gamma_5 \ell),$$

$$Q_{S_1}^\ell = \frac{e^2}{16\pi^2}(\bar{s}_L b_R)(\bar{\ell}_R \ell_L), \quad Q_{S_2}^\ell = \frac{e^2}{16\pi^2}(\bar{s}_R b_L)(\bar{\ell}_L \ell_R). \tag{4.6}$$

A key prediction of the SM is that the Wilson coefficients of the FCNC operators $Q_9^\ell$ and $Q_{10}^\ell$ are lepton-flavor universal (indeed the lepton label of these operators and the corresponding coefficients is usually omitted in SM analyses). In order to analyze NP effects that violate this prediction, it is convenient to distinguish LFU-breaking contributions from universal NP corrections to these coefficients. We choose to define the universal corrections using the Wilson coefficients of the electron modes as reference, i.e.

$$\Delta C_i^U \equiv C_i^e - C_i^{\text{SM}} \quad [i = 9, 10], \tag{4.7}$$

such that the LFU-breaking terms can be defined as

$$\Delta C_i^\mu \equiv C_i^\mu - C_i^e = C_i^\mu - (C_i^{\text{SM}} + \Delta C_i^U) \quad [i = 9, 10]. \tag{4.8}$$

(see appendix C for the numerical values of the $C_i^{\text{SM}}$). In principle, a similar decomposition can be employed for the non-standard operators in Eq. (4.6). However, in that case the SM contribution is zero. It is therefore often more convenient to label each Wilson coefficients with the lepton flavor to which it refers to (e.g. $C_9'^e$, $C_9'^\mu$, ...).

### 4.2.2 *Decay amplitudes and* clean *observables*

Following the general discussion in section 1.4.2, after having identified the $b \to s\ell^+\ell^-$ effective Lagrangian, in order to compute the decay amplitudes for the exclusive decays we are interested in we need to evaluate the matrix elements of the effective operators. For processes with at most one hadron in the final state, and considering only operators bilinear in the quark fields this is quite simple (at least if we neglect higher-order terms in the electroweak couplings). In this limit we can indeed factorise the complete matrix element into the hadronic matrix element of a quark current, and a leptonic term. For instance, in the case of the two leading FCNC operators $Q_9$ and $Q_{10}$ we can write

$$\langle \ell^+\ell^- M|Q_i|B\rangle = H_i^\mu \times L_\mu^i, \quad H_{9,10}^\mu = \langle M| \bar{s}_L\gamma^\mu b_L |B\rangle. \quad (4.9)$$

This factorization holds to all orders in QCD. It is violated by structure-dependent QED corrections, but in this case the effect is usually very small.

In order to evaluate the hadronic terms $H_i^\mu$ we need appropriate form factors, which need to be evaluated over the whole kinematical regime of the dilepton invariant mass $q^2 = (p_{\ell^+} + p_{\ell^-})^2$. For $B \to K$ and $B \to K^*$ decays these have been computed both using light-cone sum rules (LCSR) and lattice QCD. Further precision can be gained by performing combined fits of the lattice results valid at high $q^2$ and LCSR results valid at low $q^2$. Currently, the form factor uncertainties represent the dominant source of theoretical errors in predicting the dilepton spectra shown in Fig. 3.3 (see e.g. Ref. [50, 167, 174, 190, 269–271]).

Beyond the form factors, an important source of theoretical uncertainties are the matrix elements of four-quark operators, and in particular those of the leading operators $Q_{1,2}^c$ which, being generated at the tree level, have large numerical coefficients. The structure of representative hadronic matrix element for these operators are illustrated in Fig. 4.1. For $q^2$ around the masses of the $c\bar{c}$ resonances, these diagrams give rise to large contributions which cannot be estimated reliably in perturbation theory. This is why these regions are vetoed in the experimental analyses. The diagrams of the

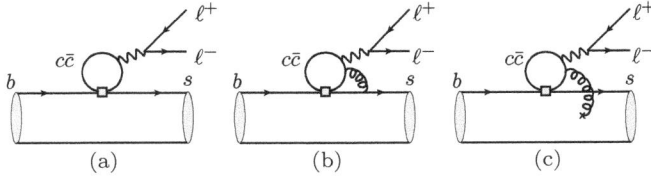

Fig. 4.1: Representative diagrams for the matrix element of the leading four-quark operators in $B \to M\ell^+\ell^-$ decays.

types (a) and (b) denote the factorizable part of the matrix element that, far from the resonance region, can be estimated reliably in perturbative QCD. The largest concern are the non-factorizable (and non-local) corrections of the type (c). These effects have been estimated in [208, 272] in the low-$q^2$ region via a power expansion in $\Lambda_{\rm QCD}/m_c$ and turn out to be quite small, in both $B \to K$ and $B \to K^*$ decays. A more general approach based on analyticity, data, and perturbative constraints [273, 274], further confirm the smallness of these non-local matrix elements. However, the uncertainty due to these non-perturbative contributions remains a source of concern due to the lack of a systematic control of all the approximations entering their evaluation. At low $q^2$, additional non-factorizable effects include weak annihilation diagrams and hard-spectator diagrams which can be estimated reliably using QCD factorization and SCET [275–279].

An important observation is that the matrix elements of four-quark operators cannot induce violations of LFU and cannot induce amplitudes where the leptonic part of the matrix element is an axial current. This allows us to define the set of so-called *clean observables*, i.e. the observables which are insensitive to the sizeable theoretical uncertainties associated to $c\bar{c}$ re-scattering (diagrams in Fig. 4.1), at least up to higher-order QED corrections. Among the quantities measured so far, this category consist of the four observables reported in Table 4.1.

- LFU ratios. The first examples of such observables are the $R_{K^{(*)}}$ ratios defined in Eq. (3.1). When considering the ratio of decay widths of identical hadronic processes into $e^+e^-$ or $\mu^+\mu^-$ pairs the only source of theoretical uncertainty, within the SM, originates from QED corrections. The latter are completely negligible if the measurements are fully inclusive with respect to the electromagnetic radiation [187, 280]. Given the present kinematical cuts (and the corresponding corrections applied by the experiments to convert the measurements into photon-inclusive

Table 4.1: Experimental results and SM predictions for the *clean observables* in $b \rightarrow s\ell^+\ell^-$ decays.

| Observable | Experiment [Tab. 3.2 & 3.5] | SM |
|---|---|---|
| $R_{K^*}^{[0.045,1.1]}$ | $0.66^{+0.11}_{-0.07} \pm 0.03$ | $0.906 \pm 0.028$ [187] |
| $R_{K^*}^{[1.1,6.0]}$ | $0.69^{+0.11}_{-0.07} \pm 0.05$ | $1.00 \pm 0.01$ [187] |
| $R_{K}^{[1.1,6.0]}$ | $0.846^{+0.042+0.013}_{-0.039-0.012}$ | $1.00 \pm 0.01$ [187] |
| $\mathcal{B}(B_s \rightarrow \mu^+\mu^-)$ | $(2.85^{+0.32}_{-0.31}) \times 10^{-9}$ | $(3.66 \pm 0.14) \times 10^{-9}$ [213] |

rates), QED corrections do not exceed 1% in $R_K$ and in the high-$q^2$ bin of $R_{K^*}$ [187, 280].

We can derive a simple general expression for LFU ratios of processes which are not helicity suppressed (such as $B \rightarrow K^{(*)}\ell^+\ell^-$, but also $B \rightarrow K\pi\ell^+\ell^-$, $\Lambda_b \rightarrow \Lambda\ell^+\ell^-$, ...), provided they are measured in dilepton invariant mass regions where both lepton masses can be neglected (i.e. for $q^2 \gg m_\mu^2$) and cutting-out the narrow charmonia resonances. Generically denoting such ratios $R_X^{(\mu/e)}$, the SM prediction is $R_{X,\mathrm{SM}}^{(\mu/e)} = 1$, and the generic expression valid beyond the SM is of the form

$$R_X^{(\mu/e)} = 1 + \sum_{i=9,9',10,10'} \eta_X^i \mathrm{Re}\,(\Delta C_i^\mu) + \mathcal{O}[(\Delta C_i^\mu)^2], \qquad (4.10)$$

where he $\mathcal{O}[(\Delta C_i^\mu)^2]$ terms can be neglected if NP is a small perturbation over the SM amplitude. The $\eta_X^i$ depend on the specific observable under consideration. However, if $q^2$ is sufficient large that the contribution of the dipole operator ($Q_7$) can be neglected, the left-handed nature of the SM amplitude implies $\eta_X^9 \approx -\eta_X^{10}$ and $\eta_X^{9'} \approx -\eta_X^{10'}$. For instance, in the case of $R_K$, with $q^2 \in [1.1,6]\,\mathrm{GeV}^2$, one gets $\eta_K^9 = \eta_K^{9'} = 0.24$ and $\eta_K^{10} = \eta_K^{10'} = -0.26$ (see appendix D).

- $\mathcal{B}(B_s \rightarrow \mu^+\mu^-)$. The conservation of angular momentum implies that the two leptons in this decay must have opposite helicity. This, in turn, implies the decay amplitude is helicity suppressed and does not receive tree-level contributions by vector-current operators, such as $Q_9$, or single-photon exchange amplitudes, such as those in Fig. 4.1. Within the SM the process is completely dominated by the contribution of $Q_{10}$ and can be estimated with high accuracy [281–283] in terms of the $B_s$ meson decay constant $f_{B_s}$

$$\langle 0|\bar{s}\gamma^\mu\gamma_5 b|B_s(p)\rangle = if_{B_s}p^\mu. \qquad (4.11)$$

A systematic analysis including higher-order QED-QCD corrections has recently been presented in [213, 284], where the overall relative theory uncertainty on $\mathcal{B}(B_s \to \mu^+\mu^-)_{\mathrm{SM}}$ is estimated to be 4%. This is due to a combination of parametric uncertainties, from $f_{B_s}$ and CKM matrix elements (leading contribution), and the recently estimated higher-order QED-QCD corrections.

In both cases, these uncertainties are well below the current experimental precision, as shown in Table 4.1. Expressions of the clean observables in terms of the SM Wilson coefficients $(C_{9,10}^\ell)$ are reported in appendix D.

$R_K$, $R_{K^*}$ and $\mathcal{B}(B_s \to \mu^+\mu^-)$ are the only clean observables measured so far, but the list is likely to increase in the near future. First of all, it is possible to conceive additional LFU ratios, such as $R_\phi$ defined in analogy to $R_{K^{(*)}}$ in $B_s \to \phi\ell^+\ell^-$. Moreover, a very interesting role is played by the so-called $Q_i = P_i^\mu - P_i^e$ defined in [189], i.e. differences in the variables describing the angular distribution in $B \to K^*\ell^+\ell^-$ [see Eq. (3.10) and Fig. 3.16] between electron and muon modes.

### 4.2.3 *New-physics hypotheses and fit to data*

The first hypothesis of NP effects of short-distance origin in $b \to s\ell^+\ell^-$ transitions related to the current anomalies was formulated in Ref. [285]. At that time only the angular distribution in $B \to K^*\mu^+\mu^-$ (and in particular the $P_5'$ observable) showed a significant deviation with respect to the SM, and the effect could be well described by a non-standard value of the Wilson coefficient $C_9$. Later on several theory groups have analysed these processes within the framework of effective Lagrangians (see e.g. Ref. [166, 188, 192, 195, 286–295]). These analyses provided fits of the coefficients under different NP hypotheses, obtaining significances that in the last few years largely exceed the $5\,\sigma$ level [293–295].

The significance of a given NP hypotheses depends both on the model we are testing against the SM, and on the treatment of the theoretical errors on the observables we are considering. As far as the NP hypotheses are concerned, it is worth stressing that there are two independent questions we would like to address: i) estimating the significance of NP in general terms, with a minimum theoretical bias; ii) identify the structure of motivated NP models describing current data. To address the first question we need to compare data against a general description of NP. The correlation of the different observables does depend on the (unknown) NP model, and only analysing all possible correlations among observables we can estimate

Table 4.2: Different NP hypotheses to describe $b \to s\ell^+\ell^-$ data and significance of the corresponding hypothesis using present data, evaluated in a conservative manner (see text).

| New Physics hypothesis | Coefficients involved | Significance [*] |
|---|---|---|
| Generic heavy NP ($b \to s\ell^+\ell^-$ contact interaction) | $C_{9,10}^{\mu},\ C_{9,10}^{\prime\mu},\ C_{S_1,S_2}^{\prime\mu}$ $C_{9,10}^{e},\ C_{9,10}^{\prime e}, C_{S_1,S_2}^{\prime e}$ | $4.3\sigma$ [296] |
| Minimally-broken $U(2)^5$ flavor symmetry | $\Delta C_9^{\mu} = -\Delta C_{10}^{\mu}$ $\Delta C_9^{\mathrm{U}},\ \Delta C_{10}^{\mathrm{U}},\ C_{S_1}^{\prime\mu}$ | $4.7\sigma$ [296] |
| $U_1$ mediator & minimally-broken $U(2)^5$ | $\Delta C_9^{\mu} = -\Delta C_{10}^{\mu}$ $\Delta C_9^{\mathrm{U}}$ | $4.8\sigma$ [300] |

correctly the probability that the pattern observed in current data is not due to a random fluctuation. On the other hand, to address the second question we need to employ specific, motivated hypotheses about the nature of physics beyond the SM. To address both these questions, in the following we restrict the attention to the following three sets of hypotheses (summarised in Table 4.2).

I. Generic heavy new physics.

This is the choice we need to employ in order to assess the significance of NP in $b \to s\ell^+\ell^-$ transitions in general terms, with minimum theoretical bias. Following Ref. [296], we can test the SM against any form of heavy NP giving rise to a short-distance effective coupling between the $b$ and the $s$ fields, and a lepton-anti-lepton pair. To do so, all the coefficients of the FCNC operators bilinear in the lepton fields must be considered as free parameters. This amount to 8 parameters for the $Q_{9,10}^{\ell}$ and $Q_{9,10}^{\prime\ell}$ operators ($\ell = e, \mu$) plus 1 effective combination of coefficients for the 4 scalar operators.[1]

II. Minimally broken $U(2)^5$ flavor symmetry.

This hypothesis is a general assumption about the flavor structure of NP which has been formulated well before the appearance of the anomalies [297–299]. The main idea is to describe in general

---

[1]The four scalar operators lead to $b \to s\ell^+\ell^-$ amplitudes which are helicity suppressed. Barring completely unrealistic models, one can thus restrict the attention to the single effective combination of Wilson coefficients which contribute to the single helicity suppressed observable measured so far, namely $\mathcal{B}(B_s \to \mu^+\mu^-)$.

terms models where new TeV-scale dynamics (possibly connected to a solution of the Higgs hierarchy problem) couples mainly to the third generation, in a manifestly flavor non-universal way. In order to satisfy the tight constraints from $B$- and $K$-meson observables, flavor mixing between the third and the light generations occurs mainly in the left-handed sector, with a CKM-like structure similar to the one observed in the quark Yukawa couplings (see appendix B for more details). When applied to the $b \to s\ell^+\ell^-$ system, this hypothesis implies that the operators affected by NP are $Q^\ell_{9,10}$ ($\ell = e, \mu$) and $Q^\mu_{S_1}$. More precisely, under this hypothesis we expect a flavor non-universal corrections to $C^\mu_9$ and $C^\mu_{10}$ of opposite sign, as well as uncorrelated lepton-universal corrections to both $C_9$ and $C_{10}$. As we shall see, within this general and motivated framework we can obtain an excellent description of present data.

III. $U(2)^5$ flavor symmetry and $U_1$ leptoquark mediator.

This last hypothesis is a restricted case of the general $U(2)^5$ framework discussed above. We supplement the ansatz about the flavor structure of the NP model with the dynamical hypothesis that the leading flavor-changing operators generated above the electroweak scale are the result of the tree-level exchange of spin-1 massive leptoquark (LQ) with the SM quantum number of an $u_R$ quark (known as the $U_1$ leptoquark). This hypothesis, which will be extensively motivated and discussed in the next sections, is favored by data if we aim at describing not only $b \to s\ell^+\ell^-$ observables, but also charged-current anomalies. As we shall clarify later on, when applied to the $b \to s\ell^+\ell^-$ system, this hypothesis implies we can limit our considerations to only two independent free parameters: a purely left-handed non-universal correction $\Delta C^\mu_L$ (affecting $C^\mu_9$ and $C^\mu_{10}$ in an opposite way), and universal correction to $C_9$ only:

$$C^e_9 = C_{9,\text{SM}} + \Delta C^U_9, \quad C^e_{10} = C_{10,\text{SM}},$$

$$C^\mu_9 = C_{9,\text{SM}} + \Delta C^U_9 + \Delta C^\mu_L, \quad C^\mu_{10} = C_{10,\text{SM}} - \Delta C^\mu_L. \tag{4.12}$$

## Conservative estimates of the NP hypothesis

A conservative estimate about the significance of the NP hypothesis can be obtained employing the first hypothesis illustrated above, estimating the sensitivity of the different observables via an ensemble of pseudo-experiments generated according to the SM hypothesis, and using the likelihood ratio as the test statistic [296]. This procedure allow us to sample all the possible correlations among observables corresponding to different NP hypotheses, and account in a rigorous way for the corresponding trial factor.

Adopting a conservative attitude toward theoretical uncertainties, we restrict the attention to the set of clean observables in Table 4.1 and the normalized angular distribution in $B \to K^* \mu^+ \mu^-$ decays. The latter is potentially sensitive to the non-local charm contributions, however, they induce additions to the decay amplitudes that can effectively be described by a $q^2$-dependent and lepton-universal shift in $C_9$:

$$\Delta C_9^U \to \Delta C_9^U + f_{B \to f}^{c\bar{c}}(q^2). \tag{4.13}$$

Hence, we can minimize the theoretical uncertainty associated with the non-local charm contributions treating $\Delta C_9^U$ as SM nuisance parameter.[2] It is useful to retain the $B \to K^* \mu^+ \mu^-$ angular distribution in this type of analysis since this distribution is sensitive to non-standard effects on FCNC operators (beyond the SM basis) even if we marginalise over $\Delta C_9^U$. The same is not true for observables such as the $q^2$ spectra.

Proceeding in this way the *global significance* of the NP hypothesis in $b \to s \ell^+ \ell^-$ transitions using current data has been estimated to be at the $4.3\sigma$ confidence level [296]. Repeating the same procedure but employing the motivated, and a priori formulated, hypothesis II illustrated above, the significance raises to $4.7\sigma$. These results provide not only a very strong evidence of NP, but also a clear indication that $b \to s \ell^+ \ell^-$ data point to a well-defined class of motivated NP models.

### Describing the anomalies in motivated frameworks

Restricting the attention to the motivated NP hypotheses II and III discussed above, we can now look more closely to the indications emerging from data.

---

[2]If we were to include more channels potentially affected by non-local charm contributions, we would need to treat the determination of $\Delta C_9^U$ from each channel as an independent nuisance parameter. In principle, charm re-scattering also naturally induces $q^2$-dependent corrections to $\Delta C_9^U$; however, these cannot be confused with a possible NP contribution which is necessarily $q^2$-independent. This is why treating $\Delta C_9^U$ as a $q^2$-independent nuisance parameter is the most conservative approach for addressing non-local charm contributions in the NP fit.

In the general $U(2)^5$ case (hypothesis II), the clean observables $R_K$ and $R_{K^*}$ are sensitive only to $\Delta C_9^\mu$ and $\Delta C_{10}^\mu$ (in the limit where we neglect small threshold effects in the low-$q^2$ bin of $R_{K^*}$), whereas $\mathcal{B}(B_s \to \mu^+\mu^-)$ is sensitive also to $\Delta C_{10}^U$ and $C_{S_1}^\mu$. Within the more restrictive LQ ansatz (hypothesis III), the clean observables are described by a single NP parameter: $\Delta C_L^\mu$.

In the left panel of Fig. 4.2 we show the result of a fit to the clean observables (blue contour lines) in the $\Delta C_9^\mu$–$\Delta C_{10}^\mu$ plane. The impact of $\mathcal{B}(B_s \to \mu^+\mu^-)$ is taken into account under the assumption that the only NP contribution to this observable is the one induced by $\Delta C_{10}^\mu$. We can test the validity of this hypothesis by looking at the following separate constraints: the parameter region preferred by $R_K$ and $R_{K^*}$ (purple), the one favored by $\mathcal{B}(B_s \to \mu^+\mu^-)$ (orange), as well as the line where $\Delta C_9^\mu = -\Delta C_{10}^\mu$. The inclusion of additional non-standard contributions to $B_s \to \mu^+\mu^-$, as expected in general terms under the hypothesis II, would amount to a translation of the $\mathcal{B}(B_s \to \mu^+\mu^-)$ bands in the vertical direction. The plot shows that the two fit regions overlap in a region compatible with $\Delta C_{10}^\mu = -\Delta C_9^\mu$, without the need for extra contributions to $\mathcal{B}(B_s \to \mu^+\mu^-)$.

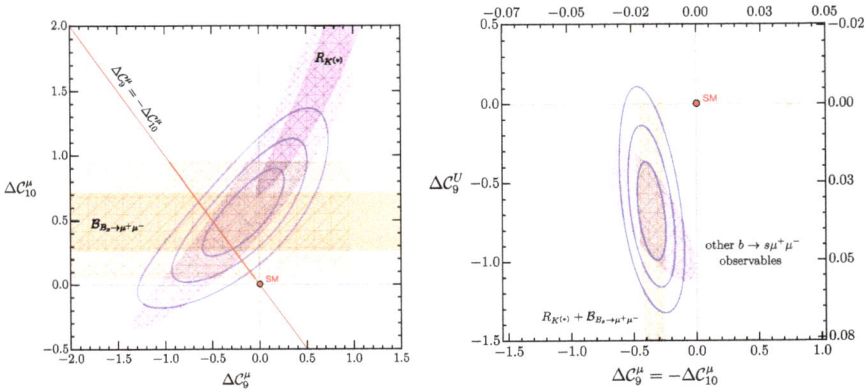

Fig. 4.2: Fits of $b \to s\ell^+\ell^-$ anomalies in motivated NP frameworks according to 2021 data (from Ref. [300]). *Left:* Results of the two-dimensional fit $\Delta C_9^\mu$ vs. $\Delta C_{10}^\mu$ using clean observables only (blue contours show $1\sigma$, $2\sigma$ and $3\sigma$ intervals). Shown in purple are the $1\sigma$ and $2\sigma$ intervals from $R_{K^{(*)}}$ and in orange $\mathcal{B}(B_s \to \mu^+\mu^-)$, the latter under the hypothesis $\Delta C_{10}^U = 0$. *Right:* Results of the two-dimensional fit $\Delta C_9^\mu = -\Delta C_{10}^\mu$ vs. $\Delta C_9^U$ using all $b \to s\ell^+\ell^-$ observables. The vertical band shows the result using clean observables only ($1\sigma$ interval), while the ellipse denote the contribution of all the other observables, estimated using the program Flavio [172] ($1\sigma$ interval). The SM prediction is shown as a red circle.

This observation provides an a posteriori consistency check that all NP effects in the clean observables can be described by a single parameter, as expected in the $U_1$ case (hypothesis III).

The $p$-value of the fit to the clean observables only, assuming $\Delta C_{10}^{\mu} = -\Delta C_9^{\mu}$ (single-parameter fit) is 12%. The significance of the NP hypothesis we are considering with respect to the SM (again based on a single-parameter fit) is $4.6\sigma$. It raises to $4.8\sigma$ if we include the contribution of the other $b \to s\ell^+\ell^-$ observables treating $\Delta C_9^U$ as a nuissance parameter. It should be stressed that this estimate of the significance is conservative, once we accept the specific, but motivated, NP hypothesis III.

In the right panel of Fig. 4.2 we show the result of a global fit of all the $b \to s\ell^+\ell^-$ observables in the $U_1$ case. The impact of the additional observables (i.e. all the $b \to s\ell^+\ell^-$ observables excluding $R_K$, $R_{K^*}$ and the $B_s \to \mu^+\mu^-$ rate) is evaluated using the public code Flavio [172]. According to the underlying NP model, we present the results as a two-dimensional fit in terms of $\Delta C_9^{\mu}$ $(= -\Delta C_{10}^{\mu})$ and $\Delta C_9^U$. As can be seen from this figure, there is a perfect consistency among the clean observables (orange band) and all the other observables (purple region). Moreover, once the strong constraint on $\Delta C_9^{\mu}$ arising from the clean observables is implemented, the other $b \to s\ell^+\ell^-$ observables can be used to constrain $\Delta C_9^U$. The latter is found to be different from zero by more than $2\sigma$. In this case the significance of the NP fit largely exceed the $5\sigma$ level; however, we stress that this result is obtained taking at face value the current best estimate the non-factorizable $c\bar{c}$ corrections (which might be affected by underestimated theory errors).

An interesting consistency check of the estimate of the non-factorizable $c\bar{c}$ corrections is provided by the extraction of $\Delta C_9^U$ in different $q^2$ bins [293]. The result of such fit using 2019 data is shown in Fig. 4.3. As can be seen, there is a good consistency among the different bins, as it is expected by a contribution of short-distance origin rather than by a mis-estimate of non-local charm contributions. The only mild tension comes from the very first bin, which reflects the too low value of $R_{K^*}^{[0.045, 1.1]}$ in Table 4.1. In the future, improved fits of this type with higher statistics could be very useful to strengthen our confidence in the theoretical control of $c\bar{c}$ contributions in $b \to s\ell^+\ell^-$.

### 4.2.4   *Summary*

Current $b \to s\ell^+\ell^-$ data provide a strong indication of physics beyond the SM. In a conservative and agnostic approach toward the nature of

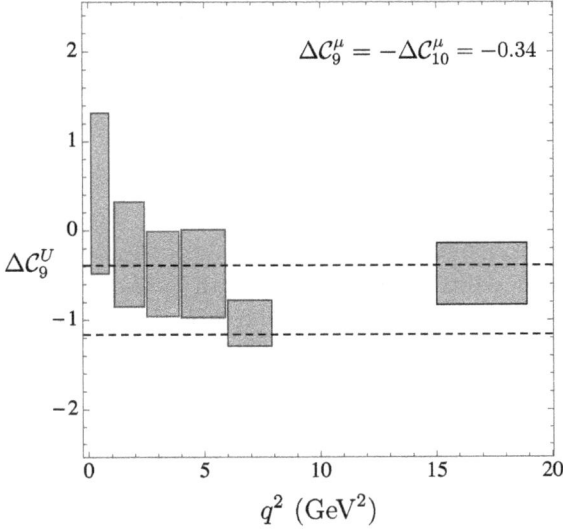

Fig. 4.3: Extraction of $\Delta C_9^U$ as a function of $q^2$ from the global fit to all $b \to s\ell^+\ell^-$ data performed in 2019 by Alguero *et al.* [293]. The results are obtained fixing $\Delta C_9^\mu = -\Delta C_{10}^\mu$ to their best fit value, as shown in the figure. The dotted line indicates the $2\sigma$ range of the ($q^2$-independent value) of $\Delta C_9^U$ from the same global fit.

NP, the significance of non-standard interactions is around the $4\sigma$ level. The significance raises to $4.6\sigma$ restricting the attention only to motivated NP frameworks, and estimating theory errors still in a very conservative manner.

Looking at $b \to s\ell^+\ell^-$ data only, it is difficult to draw definite conclusions about the nature and the energy scale of the possible new dynamics. The clear indication of a violation of LFU in left-handed currents, i.e. the evidence of $\Delta C_9^\mu = -\Delta C_{10}^\mu \neq 0$, corresponds to an effective point-like interaction

$$\Delta \mathcal{L}_{\text{LFU}}^{b \to s} = \frac{1}{\Lambda_{\text{NC}}^2}(\bar{s}_L \gamma_\mu b_L)(\mu_L \gamma^\mu \mu_L), \tag{4.14}$$

where

$$\Lambda_{\text{NC}} = \left( \frac{\sqrt{2} G_F |V_{ts}| \alpha}{\pi} |\Delta C_9^\mu| \right)^{-1/2} = (40 \pm 5)\,\text{TeV}. \tag{4.15}$$

We can denote $\Delta \mathcal{L}_{\text{LFU}}^{b \to s}$ in Eq. (4.14) as the *superweak* LFU-violating interaction in the $b \to s\ell^+\ell^-$ system. This terminology is in close analogy with the one introduced long ago by Wolfenstein [301] to describe CP violation in the $K^0$-$\bar{K}^0$ system, well before the SM was formulated. The effective scale $\Lambda_{\rm NC}$ points indeed to a four-fermion interaction which is more than four orders of magnitude weaker than ordinary weak interactions: $\Lambda_{\rm NC}^{-2}/G_F \sim 5 \times 10^{-5}$.

However, it must be stressed that the scale $\Lambda_{\rm NC}$ is a conventional scale obtained by setting to one the effecting coupling. Further hypotheses are needed in order to identify the physical scale of the new dynamics. Setting a bound on the maximal interaction strength by the request of unitarity of physical cross sections (estimated at the tree level), leads to an upper bound on the physical scale of about 80 TeV [302]. In realistic models, where the coupling strength for $b \to s\ell^+\ell^-$ amplitudes is far from being maximal, the corresponding scale of the new dynamics is much lower.

As we have seen, data fit very well with the motivated hypothesis of LQ-induced semileptonic operators respecting an approximate $U(2)^5$ flavor symmetry. In this case, which is discussed in more detail in section 4.4 and 4.6, the scale is expected to be in the few TeV domain. However, the smallness of the effective coupling in Eq. (4.14) leaves open many alternative options for its possible ultraviolet completion. This effective interaction could even be generated at the loop level, as proposed for instance in Ref. [303–307], rather than by the tree-level exchange of a single heavy mediator.

Among the tree-level mediators, beside scalar and vector leptoquarks (see section 4.5.1), an interesting alternative class of mediators are $Z'$ bosons. In particular, data would require a heavy $Z'$ boson with an effective flavor-changing coupling to quarks and a lepton-conserving (non-universal) couplings to leptons [285, 308–318]

## 4.3  The $b \to c\ell\nu$ anomalies

### 4.3.1  *Effective Lagrangian and NP hypothesis*

The effective Lagrangian describing charged-current semileptonic $B$ decays has a much simpler structure compared to the neutral current case. Focusing the attention on $b \to c$ and $b \to u$ transtions, we define the normalization of the effective operators by

$$\mathcal{L}_{b \to u^i \ell \nu} = -\frac{4G_F}{\sqrt{2}} \sum_{i=u,c} V_{ib} \sum_\alpha C_\alpha^{i\ell} Q_\alpha^{i\ell} + \text{h.c.} \qquad (4.16)$$

The complete list of $d = 6$ operators written in terms of SM fermions contributing at the tree-level to these decays is composed by five terms for each choice of quark and lepton flavors. We can conveniently spilt these terms into two sets

$$Q_{V_L}^{i\ell} = (\bar{u}_L^i \gamma_\mu b_L)(\bar{\ell}_L \gamma^\mu \nu_L^\ell), \quad Q_{S_R}^{i\ell} = (\bar{u}_L^i b_R)(\bar{\ell}_R \nu_L), \quad (4.17)$$

and

$$Q_{V_R}^{i\ell} = (\bar{u}_R^i \gamma_\mu b_R)(\bar{\ell}_L \gamma^\mu \nu_L^\ell), \quad Q_{S_L}^{i\ell} = (\bar{u}_R^i b_L)(\bar{\ell}_R \nu_L),$$
$$Q_T^{i\ell} = (\bar{u}_R^i \sigma_{\mu\nu} b_L)(\bar{\ell}_R \sigma^{\mu\nu} \nu_L), \quad (4.18)$$

where $u^i$ generically denotes the two up-type quarks (in the mass-eigenstate basis). The SM case corresponds to $C_{V_L}^{i\ell} = 1$, with the other coefficients set to zero.

According to the general hypothesis of $U(2)^5$ broken flavor symmetry, the coefficients of the set in Eq. (4.18) are all strongly suppressed by light fermion masses. Moreover, the operator $Q_{V_R}^{i\ell}$ is lepton universal unless one takes into account effects induced by $d = 8$ operators above the electroweak scale [319–321], and from a pure phenomenological point of view the effect of the scalar operator $Q_{S_R}^{i\ell}$ is very similar to that of $Q_{S_L}^{i\ell}$ (see e.g. [322]). For all these reason, in the following we restrict the attention only to the two operators in Eq. (4.17).

A further simplification concerning the basis of operators arises by observation that the precisely measured semileptonic decays into light leptons ($\ell = e, \mu$) do not show hints of non universality, see Eq. (3.14). We can thus assume that the only sizeable deviations from the SM occur in the operators involving $\tau$ leptons. Once more, this is perfectly consistent with the hypothesis of an approximate $U(2)^5$ flavor symmetry, which implies dominant NP effects in operators involving more third-generation fermions. Separating explicitly the SM contribution in the left-handed operator, we can describe NP effects in these transitions via two effective coefficients ($C_{LL}^c$ and $C_{LR}^c$) for $b \to c$ transitions, defined by

$$\mathcal{L}_{b \to u^i \tau \nu} = -\frac{4G_F}{\sqrt{2}} \sum_{i=u,c} V_{ib} \left[ \left(1 + C_{LL}^i\right)(\bar{u}_L^i \gamma_\mu b_L)(\bar{\tau}_L \gamma^\mu \nu_L) \right.$$
$$\left. -2\, C_{LR}^i (\bar{u}_L^i b_R)(\bar{\tau}_R \nu_L) \right], \quad (4.19)$$

and a corresponding set for $b \to u$ transitions. With such normalization, the LFU ratios $R_D$ and $R_{D^*}$ assume the following simple phenomenological expression (see appendix D):

$$R_D = R_D^{SM} \left[ \left| 1 + C_{V_L}^c \right|^2 + 2.6 \operatorname{Re} \left\{ \left( 1 + C_{V_L}^c \right) C_{S_R}^{c*} \right\} + 3.0 \left| C_{S_R}^c \right|^2 \right],$$

$$R_{D^*} = R_{D^*}^{SM} \left[ \left| 1 + C_{V_L}^c \right|^2 + 0.20 \operatorname{Re} \left\{ \left( 1 + C_{V_L}^c \right) C_{S_R}^{c*} \right\} + 0.12 \left| C_{S_R}^c \right|^2 \right].$$

$$(4.20)$$

As can be seen, $C_{V_L}^c$ describes a universal shift with respect to the SM amplitude. On the other hand, the scalar amplitude controlled by $C_{S_R}^{c*}$ has a very different size in the two observables. The numerical coefficients in Eq. (4.20) take into account the renormalization-group evolution of $\mathcal{L}_{b \to u^i \tau \nu}$ from a conventional high-scale $\Lambda = 2\,\mathrm{TeV}$ (see section 4.4.3).

Having factored out the CKM factor $V_{ib}$, the two sets of coefficients $\{ C_{LL}^c, C_{LR}^c \}$ and $\{ C_{LL}^u, C_{LR}^u \}$ are expected to be of similar size but not strictly correlated if we do not specify the nature of the $U(2)^5$ symmetry-breaking terms. They become strictly correlated under the (stronger) assumption of a minimal breaking of the $U(2)^5$ flavor symmetry (see appendix B). In this limit one finds [323].

$$C_{LL}^u = C_{LL}^c, \quad C_{LR}^u = C_{LR}^c. \qquad (4.21)$$

### 4.3.2   Data analysis

Several analysis of the LFU ratios $R_D$ and $R_{D^*}$, and other $b \to c$ observables, in terms of the full set of non-standard Wilson coefficients in Eq. (4.16) have been presented in the last few years [233, 248, 250, 253, 319, 324–338]. One of the most interesting results is that scalar operators alone cannot provide a good fit to data, especially concerning $R_{D^*}$: enhancing $R_{D^*}$ via scalar operators require large coefficients for these operators which are in tension with bounds from $B_c \to \tau\nu$ [333].

Once we restrict the attention to the motivated NP Lagrangian in Eq. (4.19), data other than $R_D$ and $R_{D^*}$ do not provide significant constraints on the parameter space of the effective theory [323], at least at present. Figure 4.4 shows the allowed regions for the coefficients $C_{LL}^c$ and $C_{LR}^c$ obtained from the measurements of $R_D$ and $R_{D^*}$, with the possible addition of the constraint from $\mathcal{B}(B^- \to \tau\bar{\nu})$.[3] The solid contour lines show the result of a fit to $R_D$ and $R_{D^*}$ only, while the dashed lines refer to

---

[3] We use $\mathcal{B}(B^- \to \tau\bar{\nu})_{\exp} = 1.09(24) \times 10^{-4}$ [12] and $\mathcal{B}(B^- \to \tau\bar{\nu})_{SM} = 0.812(54) \times 10^{-4}$ [153].

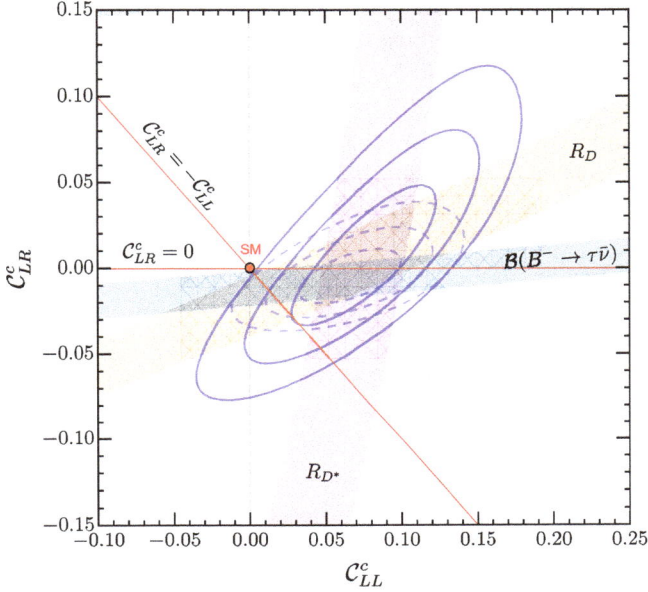

Fig. 4.4: EFT constraints from the $b \to c\tau\bar{\nu}$ anomalies in 2021 (from Ref. [300]). The solid ellipses denote the $1\sigma$, $2\sigma$ and $3\sigma$ intervals of the two-dimensional fit to $R_D$ and $R_{D^*}$ in the $C^c_{LL}$–$C^c_{LR}$ plane (coefficients evaluated at $\Lambda = 2\,\text{TeV}$). The dashed ellipses denote the fit results taking also the constraint from $\mathcal{B}(B^- \to \tau\bar{\nu})$ into account, under the hypothesis of minimal $U(2)^5$ breaking (i.e. for $C^u_{LL} = C^c_{LL}$, $C^u_{LR} = C^c_{LR}$). The bands indicate the $1\sigma$ region of each observable. The two straight lines show two benchmarks for complete models discussed in the next sections.

a fit including also $\mathcal{B}(B^- \to \tau\bar{\nu})$ under the hypothesis of minimal $U(2)^5$ breaking. In the first case, which is more conservative, the significance of the NP hypothesis compared to the SM case (two-parameter fit) is $3.2\sigma$.

The values of the Wilson coefficients reported in Fig. 4.4 are evaluated at a high renormalization scale: $\Lambda = 2\,\text{TeV}$. While the renormalization effects are negligible for $C^c_{LL}$, the mixed-chirality coefficients $C^c_{LR}$ exhibits a sizable scale variation due to QCD corrections. In Fig. 4.4 we also show as red lines the relations $C^c_{LR} = 0$ and $C^c_{LR} = -C^c_{LL}$ which are two illustrative benchmarks for specific UV completions of the effective theory we are considering. Without the inclusion of $\mathcal{B}(B^- \to \tau\bar{\nu})$, present data are compatible with both benchmarks: $C^c_{LR} = 0$ and $C^c_{LR} = -C^c_{LL}$, and likewise with any intermediate case. The inclusion of $\mathcal{B}(B^- \to \tau\bar{\nu})$ under the hypothesis of minimal $U(2)^5$ breaking slightly disfavors (by less than $2\sigma$) the scenario $C^c_{LR} = -C^c_{LL}$.

Whereas data other than $R_D$ and $R_{D^*}$ do not provide significant constraints at present, in principle other observables could be used in the future to over constrain and test the NP interpretation by means of the effective Lagrangian in Eq. (4.19). Note in particular that the left-handed operator would simply re-scale the SM amplitude, leading to a universal correction in $R_D$ and $R_{D^*}$ (relative to the corresponding SM predictions) and no change in the polarization asymmetries,

$$F_L^{D^*} = \frac{\Gamma(\bar{B} \to D_L^* \tau \bar{\nu})}{\Gamma(\bar{B} \to D^* \tau \bar{\nu})}, \quad P_\tau^{D^{(*)}} = \frac{\Gamma(\bar{B} \to D^{(*)} \tau^+ \bar{\nu}) - \Gamma(\bar{B} \to D^{(*)} \tau^- \bar{\nu})}{\Gamma(\bar{B} \to D^{(*)} \tau^+ \bar{\nu}) + \Gamma(\bar{B} \to D^{(*)} \tau^- \bar{\nu})}.$$

$$(4.22)$$

On the other hand, these differ from the SM case if $C_{LR}^c = 0$. One therefore expects a strong correlation between the possible deviations from the SM in the polarization asymmetries and the difference of NP effects in $R_D$ and $R_{D^*}$. This is illustrated in Fig. 4.5, where we plot the deviations from unity

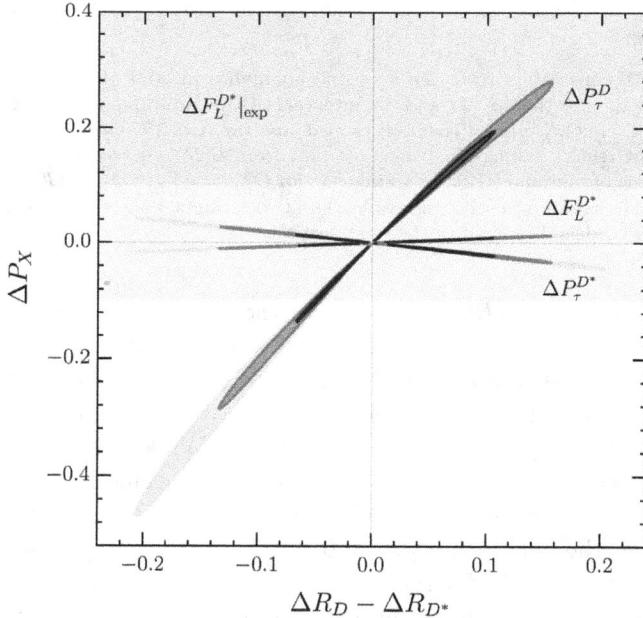

Fig. 4.5: Deviations of the polarization asymmetries compared to the SM as a function of $\Delta R_D - \Delta R_{D^*}$, in the generic NP framework described by the effective Lagrangian in Eq. (4.19). The predictions are obtained using the fit in Fig. 4.4 (continuous lines). The horizontal gray bands indicate, the experimental value of $\Delta F_L^{D^*}$ at $1\sigma$ and $2\sigma$ (figure from Ref. [323]).

of the polarization ratios vs. the difference on the two leading LFU ratios, defined by

$$\Delta P_X = \frac{P_X}{P_X^{SM}} - 1, \quad \Delta R_X = \frac{R_X}{R_X^{SM}} - 1. \tag{4.23}$$

As can be seen, the predicted pattern of deviations is very precise and rather specific. At present, only $P_\tau^{D^*}$ [243, 339] and $F_L^{D^*}$ [255] have been measured (see section. 3.4.2), but the results are still affected by large uncertainties and do not add significant information.

### 4.3.3 *Summary*

The significance of the NP hypothesis is $b \to c\tau\bar{\nu}$ amplitudes is slightly above $3\sigma$. It is not a high significance if taken alone, but is quite interesting being a second independent hint of LFU violation in semileptonic $B$ decays. Data are compatible with a pure left-handed interaction (i.e. a simple overall rescaling of the SM amplitude), but sizeable scalar (right-handed) amplitudes cannot be excluded at this stage. The evidence of LFU in left-handed currents, corresponds to an effective point-like interaction

$$\Delta \mathcal{L}_{LFU}^{b\to c} = -\frac{1}{\Lambda_{CC}^2}(\bar{c}_L\gamma_\mu b_L)(\tau_L\gamma^\mu\nu_L^\tau), \tag{4.24}$$

where

$$\Lambda_{CC} = \left(\frac{4G_F|V_{cb}|}{\sqrt{2}}|C_{LL}^c|\right)^{-1/2} = (3.5 \pm 0.8)\,\text{TeV}. \tag{4.25}$$

Similarly to $\Lambda_{NC}$ in Eq. ( 4.15), also $\Lambda_{CC}$ is a pure conventional scale. In this case the unitarity bound on the NP scale is about 9 TeV [302]. Such a low effective scale points to tree-level mediators for this effective interaction.

## 4.4 Combined analysis of the two anomalies

In this section we analyse the possibility of addressing both set of anomalies within a common EFT approach. The main idea is to consider the apparently different values of $\Lambda_{NC}$ and $\Lambda_{CC}$ as the result of the non-trivial flavor structure on the same underlying dynamics. This possibility is clearly more speculative; however, it is also more fascinating and challenging from the theoretical point of view. As we shall see, this hypothesis leads to a rather constrained NP construction, pointing not only to a motivated flavor structure, but also to a well-defined class of mediators. This implies a series of predictions for processes other than $b \to s\ell^+\ell^-$ and $b \to c\tau\bar{\nu}$, at both low- and high-energies, which can be tested or falsified in the near future.

Before starting this combined EFT analysis, it is worth recalling the key steps performed over the last few years in analysing and interpreting both sets of anomalies by means of a common theoretical framework.

### 4.4.1  *Historical remarks and main strategy*

- *Left-handed nature of the LFU-violating interactions.* The first important observation has been the identification of a purely left-handed contact interaction, involving muons only, as the key ingredient to explain $b \to s\ell^+\ell^-$ data. As we already stated, the first hypothesis of NP effects of short-distance origin in $b \to s\ell^+\ell^-$ transitions related to the current anomalies was formulated in Ref. [285] in 2013, when only the angular distribution in $B \to K^*\mu^+\mu^-$ showed a significant deviation with respect to the SM, and data were best described by a non-standard effect in $C_9$ only. In 2014, when the anomaly in $R_K$ also appeared, Hiller and Schmalz pointed out that a purely left-handed contact interaction was a more motivated choice from the model-building point of view [288]. As we already discussed, later on global analyses of $b \to s\ell^+\ell^-$ data have shown that indeed a purely left-handed interaction involving muons only is a necessary ingredient to explain $b \to s\ell^+\ell^-$ data.

  Electroweak symmetry unavoidably implies a link between neutral- and charged-current interactions for left-handed fields. This is why soon after the hypothesis of a left-handed interaction in the neutral-current case was formulated, it was realised that both set of anomalies could be linked via a common EFT approach involving left-handed fermions [319, 340–342].

- *Flavor structure.* The second important observation, which also started to emerge in 2014–2015, is the peculiar flavor structure of the new interaction. Glashow, Guadagnoli and Kane [343] proposed that the super-weak interaction involving muons only explaining $B \to K\ell^+\ell^-$ data could be the result of a stronger interaction, violating lepton flavor, involving mainly third-generation fermions [343]. But the crucial step foward, for the combination of the two set of anomalies, has been the observation made in Ref. [341, 344] (see also [345]) that the relative strength of the two interactions is well compatible with the hypothesis of a minimally broken $U(2)^5$ flavor symmetry. As already stressed, this hypothesis was formulated well before the anomalies appeared [297–299], in order to describe in general terms models with new dynamics at the

TeV scale, coupled mainly to the third generation, respecting both flavor and collider bounds. The link between the $B$ anomalies and the $U(2)^5$ flavor symmetry opened the interesting possibility of an explanation of the anomalies in terms of some underlying dynamics responsible also for the hierarchies observed in the SM Yukawa couplings.

- *Leptoquarks.* Heavy leptoquark fields soon appeared as the most efficient mediators to describe the semileptonic contact interactions observed at low energies. This already holds when considering $b \rightarrow s\ell^+\ell^-$ transitions only [288, 346, 347], and even more in the case of combined explanations [319, 325, 342, 344, 348–350]. The phenomenological success of the LQ hypothesis in explaining the anomalies is quite simple: leptoquarks do contribute at the tree-level in these transitions, which do exhibit anomalies, while they contribute only at the loop level in four-quark or four-lepton contact interactions, which so far do not exhibit deviations from the SM. LQ fields are present in a wide class of SM extensions aiming at a unified description of quark and lepton quantum numbers (see e.g. [351–353]). They are also present, as composite states, in models featuring new strong dynamics above the electroweak scale. Actually in this context the possibility of a LQ-induced effect in $b \rightarrow s\mu^+\mu^-$ transitions was proposed by Gripaios well before the anomalies appeared [354].

- *High-energy constraints.* The last important step for models addressing both set of anomalies, is the relevance of bounds beyond flavor physics, from collider searches at high energies and electroweak precision observables. At the end of 2015 various options for a combined explanation of the anomalies were open, including hypotheses based on colorless mediators (such as $W'$ and $Z'$ fields). However, detailed analyses of electroweak precision observables [355, 356], and high-$p_T$ bounds [357, 358] showed that only few options could survive these bounds. As clearly demonstarted by the EFT analysis presented in [359], which analysed for the first time all these constraints in a systematic way, only two LQ-simplified models basically survived. Among them, there is only one option with a single heavy mediator that does not require a significant tuning: a TeV-scale vector leptoquark with SM gauge quantum numbers $(\mathbf{3}, \mathbf{1})_{2/3}$, often referred to as the $U_1$ field.

The simplified model proposed by Barbieri *et al.* at the end of 2015 [344] was the first proposal based on the two key ingredients illustrated above: the $U_1$ leptoquark and the $U(2)^5$ flavor symmetry. So far, this setup still represents the most interesting attempt to explain the two sets of anomalies in a framework which does not appear to be *ad hoc* (or highly tuned) and which points toward motivated high-energy dynamics. A massive vector field necessarily requires a UV completion and, as pointed out in [344], the $U_1$ field naturally points toward the $SU(4)$ group unifying quarks and leptons, which was proposed by Pati and Salam in 1974 [360]. This is why we will dedicate more attention to this option in the following sections. In particular, in section 4.6 we address both low- and high-energy phenomenology of the $U_1$ field, while in section 4.7 we discuss in more detail some of the possible UV completions.

### 4.4.2 *Operator basis*

In order to perform a combined EFT analysis of the two anomalies we need to define a consistent operator basis containing all possible relevant terms. Rather than starting from the most general basis of semileptonic operators, we can take advantage of the studies performed in the last few years, summarised above, and concentrate on a relatively limited set. Once all constraints (including high-$p_T$ data, electroweak precision tests and other flavor observables) are taken into account, the basis of semileptonic effective operators relevant to the two anomalies, at a scale $\Lambda \gg v$, reduces to the following simple set [300]:

$$
\begin{aligned}
\mathcal{O}_{LL}^{ij\alpha\beta} &= (\bar{q}_L^i \gamma_\mu \ell_L^\alpha)(\bar{\ell}_L^\beta \gamma^\mu q_L^j) = \frac{1}{2}\left[Q_{lq}^{(1)} + Q_{lq}^{(3)}\right]^{\beta\alpha ij}, \\
\mathcal{O}_{LR}^{ij\alpha\beta} &= (\bar{q}_L^i \gamma_\mu \ell_L^\alpha)(\bar{e}_R^\beta \gamma^\mu d_R^j) = -2[Q_{ledq}^\dagger]^{\beta\alpha ij}, \\
\mathcal{O}_{RR}^{ij\alpha\beta} &= (\bar{d}_R^i \gamma_\mu e_R^\alpha)(\bar{e}_R^\beta \gamma^\mu d_R^j) = [Q_{ed}]^{\beta\alpha ij},
\end{aligned}
\tag{4.26}
$$

where a contraction of color and $SU(2)_L$ indices between fermions inside parenthesis is implied. It should be stressed that the operators above are not the complete set of operators relevant at low energies, but only the set of presumably dominant terms at high energies due to the new dynamics. Their renormalization group (RG) evolution down to low energies necessarily generates many additional terms which can (and will be) analysed in a systematic way.

Here $Q_{lq}^{(1,3)}$, $Q_{ledq}$ and $Q_{ed}$ denote the operators in the the so-called Warsaw basis of the SMEFT [58] (see section 1.4.4). In the Warsaw basis, the semileptonic operators are written as product of a quark current times a leptonic current. This is not the case for the "leptoquark currents" appearing on the left-hand side of Eq. (4.26). The two sets are related by Fierz identities, as shown in Eq. (4.26). In the case of $Q_{lq}^{(1,3)}$, which involves left-handed currents, the superscript distinguishes the case of $SU(2)_L$-singlets or $SU(2)_L$-triplet currents.

We write the effective Lagrangian describing the NP contributions as

$$\mathcal{L}_{\text{EFT}}^{\text{NP}} = -\frac{2}{v^2} \left[ C_{LL}^{ij\alpha\beta} \, \mathcal{O}_{LL}^{ij\alpha\beta} + C_{RR}^{ij\alpha\beta} \, \mathcal{O}_{RR}^{ij\alpha\beta} + \left( C_{LR}^{ij\alpha\beta} \, \mathcal{O}_{LR}^{ij\alpha\beta} + \text{h.c.} \right) \right],$$

$$(4.27)$$

i.e. using the Fermi scale as overall normalization (we recall that $2/v^2 = 4G_F/\sqrt{2}$) and absorbing the dependence of the NP scale ($\Lambda$) inside the Wilson coefficients. The coefficients of the hermitian operators $\mathcal{O}_{LL}$ and $\mathcal{O}_{RR}$ satisfy $C_{LL}^{ji\beta\alpha} = \left( C_{LL}^{ij\alpha\beta} \right)^*$ and $C_{RR}^{ji\beta\alpha} = \left( C_{RR}^{ij\alpha\beta} \right)^*$.

We assume that these coefficients respect an approximate $U(2)^5$ flavor symmetry, with non-negligible breaking terms only in the left-handed sector (on both the quarks and leptons). This assumption implies that the leading couplings in $\mathcal{L}_{\text{EFT}}^{\text{NP}}$ are those with third-generation indices, while all other couplings are suppressed. More specifically, we make the following two assumptions:

- Wilson coefficients of operators containing first- or second-generation right-handed fields are negligibly small, i.e. $C_{RR}^{ij\alpha\beta} \approx 0$ unless $i = j = 3$ and $\alpha = \beta = \tau$, and $C_{LR}^{ij\alpha\beta} \approx 0$ unless $j = 3$ and $\beta = \tau$.
- Wilson coefficients associated with second-generation left-handed particles are suppressed (relative to those for third-generation particles) by factors of $\epsilon_q, \epsilon_\ell \sim 10^{-1}$ for each second-generation quark or lepton, e.g. $C_{LL}^{23\tau\tau} \sim \epsilon_q \, C_{LL}^{33\tau\tau}$, $C_{LL}^{23\mu\mu} \sim \epsilon_q \, \epsilon_\ell^2 \, C_{LL}^{33\tau\tau}$ etc., and a further suppression arises in the case of operators involving first-generation fields.

As explicitly indicated by the labels, the flavor basis of the lepton fields is taken to be the charged-lepton mass basis. Unless otherwise specified, the flavor basis of the quark fields is identified with the mass basis of the down-type quarks. In such bases, the relation between the lepton doublets and the mass-eigenstates of the charged fermions ($u^i$, $d^i$ and $\ell^i$) is given by

$$q_L^i = \begin{pmatrix} V_{ji}^* u_L^j \\ d_L^i \end{pmatrix}, \quad \ell_L^i = \begin{pmatrix} \nu_L^i \\ e_L^i \end{pmatrix}. \qquad (4.28)$$

Note that a change of basis from down-type to up-type quarks would not invalidate the scaling discussed above, and would suggest that, at least in the quark sector, first-generation indices bring an additional $\epsilon_q$ suppression compared to second-generation indices. The flavor structure specified by the two assumptions stated above is the rationale behind the combined explanation of the two sets of anomalies and their possible connection to the dynamics underlying the structure of the SM Yukawa matrices. As we shall show, these scaling rules are clearly supported by the present data.

The operator basis in Eq. (4.27) corresponds to the operators generated by the tree-level exchange of the $U_1$ leptoquark. As pointed out in [323], one can obtain the same set of operators and flavor structure by starting from the full set of semileptonic SMEFT operators and imposing the assumption of a minimally-broken $U(2)^5$ flavor symmetry, without any hypothesis about the mediator. The only relevant difference under this more general hypothesis is that the operators $Q_{lq}^{(1)}$ and $Q_{lq}^{(3)}$ can appear in a different linear combination than in Eq. (4.26). An EFT analysis leaving their coefficients as free parameters has been performed in [359], where it was shown that the combination orthogonal to $\mathcal{O}_{LL}$ is tightly constrained by data on $b \to s\bar{\nu}\nu$ transitions and electroweak precision tests (at least for the leading flavor structures). Since this combination is severely constrained, and it does not lead to a qualitative change in the description of the two sets of anomalies, we shall not consider it further in this section.

### 4.4.3 *Combined EFT analysis*

#### Matching conditions

Matching the high-scale Lagrangian $\mathcal{L}_{\mathrm{EFT}}^{\mathrm{NP}}$ in Eq. (4.27) to the low-energy Lagrangians in Eqs. (4.1) and (4.16), we can derive relations between the (low-energy) coefficients determined by data and the high-scale parameters.

Starting from the non-universal correction term in $b \to s\ell^+\ell^-$ amplitudes, we find

$$\Delta C_9^\mu = -\Delta C_{10}^\mu = -\frac{2\pi}{\alpha V_{ts}^* V_{tb}} \left[ C_{LL}^{23\mu\mu} - C_{LL}^{23ee} \right] \approx -\frac{2\pi}{\alpha V_{ts}^* V_{tb}} C_{LL}^{23\mu\mu}. \quad (4.29)$$

The last relation follows from the assumption that $|C_{LL}^{23ee}| \ll |C_{LL}^{23\mu\mu}|$. The multiplicative correction of this term due to the RG evolution from the NP scale $\Lambda = \mathcal{O}(\text{few TeV})$ to the electroweak scale is at the percent level and can be safely neglected. As a result, the LFU-violating corrections $\Delta C_{9,10}^\mu$ are, to a good accuracy, scale independent.

On the other hand, since $|C_{LL}^{23\tau\tau}| \gg |C_{LL}^{23\mu\mu}|$, the mixing of $\mathcal{O}_{LL}^{23\tau\tau}$ into operators containing light leptons is an important effect. As pointed out by Crivellin *et al.* [361] (see also [362, 363]), this implies a further connection between charged-current and neutral-currents observables: the operator $\mathcal{O}_{LL}^{23\tau\tau}$, which describes tree-level contributions to $b \to c\tau\bar{\nu}$ amplitudes, is also responsible, via RG effects, of the lepton-universal corrections in $b \to s\ell^+\ell^-$ amplitudes. More precisely, the following loop-induced contributions to $C_9^U$ is generated:

$$\Delta C_9^U(m_b) = \frac{1}{V_{ts}^* V_{tb}} \frac{2}{3} \sum_{\ell=e,\mu,\tau} C_{LL}^{23\ell\ell}(\Lambda) \ln \frac{\Lambda^2}{m_b^2}$$

$$\approx \frac{1}{V_{ts}^* V_{tb}} \frac{2}{3} C_{LL}^{23\tau\tau}(\Lambda) \ln \frac{\Lambda^2}{m_b^2}. \tag{4.30}$$

Since we start from semileptonic operators at the high scale, the lepton-universal correction to $C_{10}$ turns out to be suppressed by the square of the $\tau$-lepton Yukawa coupling and hence safely negligible.

Concerning the effective coefficients relevant to charged current defined in Eq. (4.19), we find

$$C_{LL}^i \equiv C_{LL}^{33\tau\tau} \left[1 + \frac{C_{LL}^{23\tau\tau}}{C_{LL}^{33\tau\tau}} \frac{V_{is}}{V_{ib}} \left(1 + \frac{C_{LL}^{13\tau\tau}}{C_{LL}^{23\tau\tau}} \frac{V_{ud}}{V_{is}}\right)\right], \tag{4.31}$$

and similarly for $C_{LR}^{u_i}$ with $LL \to LR$ everywhere. Similarly to $\Delta C_9^\mu$, the multiplicative RG evolution of $C_{LL}^i$ is small and can be safely neglected. On the other hand, $C_{LR}^{u_i}$ undergoes large RG effects due to QCD correcrtions. The value reported in Fig. 4.4 corresponds to the high-scale value of the coupling obtained for $\Lambda = 2\,\text{TeV}$.

## Combined analysis of charged- and neutral-current data

Among all the operators (and coefficients) appearing in the high-scale Lagrangian $\mathcal{L}_{\text{EFT}}^{\text{NP}}$ in Eq. (4.27), those entering the two sets of anomalies are summarised in Table 4.3. The most precisely determined coupling is $C_{LL}^{23\mu\mu}$, which controls the breaking of universality in $b \to s\ell^+\ell^-$ decays. Its tiny but clearly non-vanishing value is five orders of magnitude smaller compared to the Fermi coupling. Significantly stronger interactions are indicated by the other couplings in Table 4.3. The leading couplings in the last row, involving only the third generation fields, are two orders of magnitudes smaller compared to the Fermi coupling, indicating new dynamics at the TeV scale.

In Fig. 4.6 we analyse in more detail the interplay of the the couplings $C_{LL}^{33\tau\tau}$ and $C_{LL}^{23\tau\tau}$ which appear in both sets of anomalies. The constraint from $R_D$ and $R_{D^*}$ are obtained in the limit where the presumably small

Table 4.3: Coefficients of the high-scale effective Lagrangian $\mathcal{L}_{\mathrm{EFT}}^{\mathrm{NP}}$ in Eq. (4.27) which provides a good description of both sets of anomalies. In the last two columns we indicates the plot which illustrate the constraints on these couplings, and the corresponding statistical significance.

| Coefficients | Anomalies | Size | Plot | Sig. |
|---|---|---|---|---|
| $C_{LL}^{23\mu\mu}$ | $(b \to s\ell^+\ell^-)_{\mathrm{clean}}$ | $-(2.0 \pm 0.5) \times 10^{-5}$ | Fig. 4.2$_{[\text{via Eq. (4.29)}]}$ | $>4\sigma$ |
| $C_{LL}^{23\tau\tau}$ | $b \to s\ell^+\ell^-,\ b \to c\tau\bar{\nu}$ | $\sim 10^{-3}$ | Fig. 4.6 | $\sim 3\sigma$ |
| $C_{LL}^{33\tau\tau},\ C_{LR}^{33\tau\tau}$ | $b \to c\tau\bar{\nu}$ | $\sim 10^{-2}$ | Fig. 4.6 | $\sim 3\sigma$ |

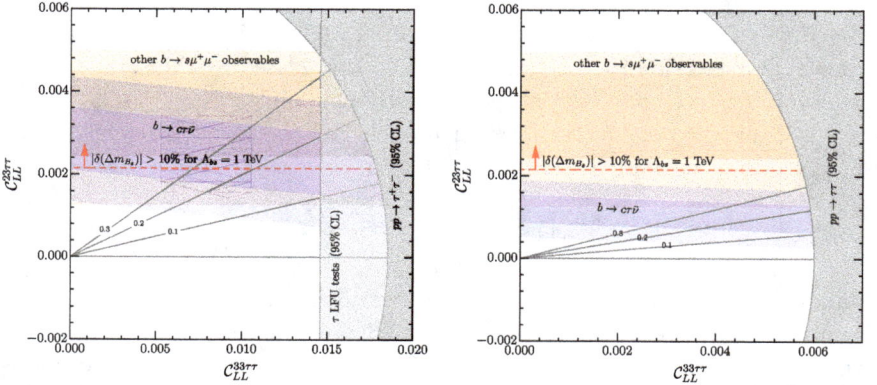

Fig. 4.6: Combined analysis of the coefficients $C_{LL}^{33\tau\tau}$ and $C_{LL}^{23\tau\tau}$ from the two set of anomalies (from Ref. [300]). The two panels correspond to the two benchmarks defined in Eq. (4.32), with the case $C_{LR}^c = 0$ on the left. The blue bands denote the $1\sigma$ and $2\sigma$ regions preferred by $b \to c\tau\bar{\nu}$ data, while the gray bands show the exclusion regions derived from $\sigma(pp \to \tau^+\tau^- + X)$. The preferred values of $C_{LL}^{23\tau\tau}$ derived from $b \to s\mu^+\mu^-$ data (at $1\sigma$ and $2\sigma$) are indicated by the horizontal orange bands. The dashed red lines provide a qualitative indication of the bound from $B_s - \bar{B}_s$ mixing (the regions above the lines are disfavoured, see text for more details). The grey lines indicate reference values of the ratio $C_{LL}^{23\tau\tau}/C_{LL}^{33\tau\tau} \sim \epsilon_q$.

contribution to $C_{LL}^c$ proportional to $C_{LL}^{13\tau\tau}$ in Eq. (4.31) is neglected. The blue band in Fig. 4.6 shows the allowed $1\sigma$ and $2\sigma$ regions in the $C_{LL}^{33\tau\tau} - C_{LL}^{23\tau\tau}$ plane for the following two benchmarks:

$$\mathbf{I.}\ C_{LR}^{i3\tau\tau} = 0 \to C_{LR}^c = 0,$$

$$\mathbf{II.}\ C_{LR}^{i3\tau\tau}(\Lambda) = -C_{LL}^{i3\tau\tau}(\Lambda) \to C_{LR}^c = -C_{LL}^c. \tag{4.32}$$

The horizontal orange bands shows the constraint on $C_{LL}^{23\tau\tau}$ from $b \to s\mu^+\mu^-$ observables, via $\Delta C_9^U$, after we marginalize over $\Delta C_9^\mu$ (i.e. after we integrate over all allowed values of $\Delta C_9^\mu$, according to the results shown in the right panel of Fig. 4.2). As can be seen, there is an excellent overlap of the two bands, which is a highly non-trivial result.

## Constraints beside the anomalies

As already pointed out in section 4.4.1, the key observation that makes leptoquark mediators particulary motivated in describing the $B$ anomalies is that they do not generate tree-level contributions to four-quark and four-lepton amplitudes, which are strongly bounded by data (in particularly via meson-antimeson mixing and leptonic $\tau$ decays). These amplitudes are generated beyond the tree level, and the generic EFT approach we are considering in this section does not allow us to address them in a very precise way. A similar argument holds for the bounds from high-energy observables, where the nature of the mediator is relevant. These bounds will be analyzed in a more systematic way in the context of an explicit model in the next section.

However, using the EFT we can still provide an instructive semi-quantitative estimate of the most relevant constraints beside the anomalies. As shown in Fig. 4.6, these are derived from: (i) high-energy measurements of the $\tau^+\tau^-$ production cross section at the LHC; (ii) tests of universality in leptonic $\tau$ decays; (iii) the $B_s$–$\bar{B}_s$ mixing amplitude.

The $pp \to \tau^+\tau^- + X$ processes is modified by the semileptonic contact interactions in Eq. (4.27) and provides relevant bounds, as pointed out first in Ref. [357]. Even if we assume these interactions are not local at high energies, the corresponding bounds from the LHC are weakly sensitive to the details of the UV completion of the EFT: at the energies presently accessible, the effect of the heavy (multi-TeV scale) mediators is well described by a contact interaction. The bounds reported in Fig. 4.6 take into account the contributions from $b\bar{b}$-, $b\bar{s}$- and $s\bar{b}$-initiated scattering processes [300]. For small $C_{LL}^{23\tau\tau}$ values, the leading contribution is due to $b\bar{b} \to \tau\tau$, but for higher values of $C_{LL}^{23\tau\tau}$ also the $s(\bar{s})$ contribution starts to be relevant. This explains the curved shape of this bound in Fig. 4.6.

Also the bounds from precision measurements of leptonic $\tau$ decays are mildly affected by the UV completion of the EFT. Purely leptonic processes are not affected at the tree-level by the semileptonic Lagrangian in Eq. (4.27). However, they receive contributions beyond the tree level providing relevant bounds, as pointed out first in [355, 356]. A representative one-loop diagram is shown in Fig. 4.7. The contribution is logarithmically

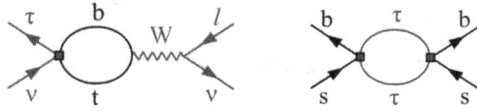

Fig. 4.7: Representative diagrams indicating the non-standard contributions to leptonic $\tau$ decays and $B_s$–$\bar{B}_s$ mixing generated beyond the tree level by the semileptonic operators describing the anomalies (indicated by the small red vertices).

divergent, indicating the need of a UV completion of the EFT. However, the leading effect is the RG mixing of semileptonic operators into leptonic operators, which can be estimated in a model-independent way. The bound reported in Fig. 4.6 (left) corresponds to a RG evolution from $\Lambda = 2\,\mathrm{TeV}$.

The EFT treatment of $B_s$–$\bar{B}_s$ mixing is the most model dependent. The relevant one-loop diagram is shown in Fig. 4.7 (right). This diagram is quadratically UV-divergent, indicating a high sensitivity to UV dynamics. For instance, in models where the $U_1$ leptoquark is a massive gauge boson [364, 365], the effective cut-off of this amplitude ($\Lambda_{bs}$) is closely connected to the mass of the vector-like leptons which are present in the models. The fact that in these models $\Lambda_{bs}$ is associated with the mass of a colorless particle is a welcome (and highly non-trivial) feature, which allows for a relatively low value of $\Lambda_{bs} \lesssim 1\,\mathrm{TeV}$ without conflicting with current LHC bounds. The horizontal red line in Fig. 4.6 corresponds to $\delta(\Delta m_{B_s}) = 0.1$, i.e. to a 10% correction to the magnitude of $\Delta m_{B_s}$, under the assumption $\Lambda_{bs} = 1\,\mathrm{TeV}$. Without entering into model-dependent considerations, we note that the absence of direct signals of new physics at the energy frontier implies that is very hard to conceive explicit models with $\Lambda_{bs}$ much below 1 TeV. This is why the region above the red line should be considered as disfavored by $B_s$–$\bar{B}_s$ mixing, barring (unnatural) cancellations with additional UV contributions, controlled by other parameters.

### 4.4.4  *Summary*

From the combined analysis of the two anomalies summarised in Fig. 4.6 and Table 4.3, we deduce that the effective Lagrangian $\mathcal{L}_{\mathrm{EFT}}^{\mathrm{NP}}$ in Eq. (4.27) provides a very good description of present data. This is true also after taking into account the tight constraints from purely leptonic and pure hadronic low-energy processes non related to the anomalies, as well as the absence of direct signals of NP at high energies. Moreover, data are perfectly consistent with the flavor scaling of the different couplings in $\mathcal{L}_{\text{EFT}}^{\text{NP}}$ presented in section 4.4.2.

An important point to stress is that the updated bounds from $pp \to \tau^+\tau^-$ prevent a solution of the $b \to c\tau\bar{\nu}$ anomalies with vanishing $C_{LL,LR}^{23\tau\tau}$, which was possible until a few years ago. This enhances the compatibility of the two anomalies, as shown in Fig. 4.6. At the same time, $B_s$–$\bar{B}_s$ mixing tends to favor the smallest possible value of $|C_{LL}^{23\tau\tau}|$, favoring the consistency of the ratio $C_{LL}^{23\tau\tau}/C_{LL}^{33\tau\tau}$ with the hierachical structure dictated by the hypothesis of a minimal breaking of the $U(2)^5$ flavor symmetry. Finally, it is worth noting that while the benchmark scenario with $C_{LR}^{i3\tau\tau} = 0$ appears to be favored by the $b \to c\tau\bar{\nu}$ data alone (see Fig. 4.4), the $B_s$–$\bar{B}_s$ mixing bound is more stringent in this case (see Fig. 4.6). Hence, with present data, it is still useful to consider both possibilities, keeping in mind that any intermediate case could also be a possible solution.

As anticipated, this EFT construction strongly points toward tree-level leptoquark mediators, coupled mainly to third generation fermions, with mass in the few TeV regime:

$$M_{\text{LQ}} \sim \frac{g_{\text{LQ}} M_W}{g_2 \sqrt{|C_{LL}^{33\tau\tau}|}} \approx (4 \pm 1)\,\text{TeV} \times \left(\frac{g_{\text{LQ}}}{3}\right). \tag{4.33}$$

## 4.5 High-scale mediators: general considerations

In this section we briefly comment on the different possibilities of generating the semileptonic contact interactions summarised in Table 4.3, which describe the two sets of anomalies, as a result of integrating out (at the tree level) some heavy mediator. This step is usually referred to as the identification of a *simplified model* for the anomalies. As the term says: the aim is not to derive a complete model, but only the leading mediator describing the anomalies. The identification of a simplified model is essential to derive precise prediction for high-energy observables, and of course it represents a first important step toward the identification of UV consistent model.

### 4.5.1 *Leptoquarks*

As anticipated in section 4.4.1, leptoquark fields soon appeared to be the most suitable candidates to explain the two anomalies (see in particular Ref. [288, 319, 325, 342, 344, 346–350, 366, 367]). This is mainly because leptoquarks do contribute at the tree-level to semileptonic processes,

Table 4.4: Summary of the LQ models which can accommodate $R_{K^{(*)}}$ (first column), $R_{D^{(*)}}$ (second column), and both $R_{K^{(*)}}$ and $R_{D^{(*)}}$ (third column). Only the $U_1$ leptoquark model doesn't have phenomenological problems, which are listed for the other models (from Ref. [368]).

| Leptoquark field | $R_{K^{(*)}}$ | $R_{D^{(*)}}$ | Main problem |
|---|---|---|---|
| $S_1 = (\bar{\mathbf{3}}, \mathbf{1})_{1/3}$ | ✗ | ✓ | $[R_{K^{(*)}}]_{\mathrm{loop}}$ vs. $B$ decays & EWPO |
| $R_2 = (\mathbf{3}, \mathbf{2})_{7/6}$ | ✗ | ✓ | $[R_{K^{(*)}}]_{\mathrm{loop}}$ vs. high-$p_T$ bounds |
| $\widetilde{R}_2 = (\mathbf{3}, \mathbf{2})_{1/6}$ | ✗ | ✗ | $R_{K^*} < R_{K^*}^{\mathrm{SM}}$, $R_{D^{(*)}} \approx R_{D^{(*)}}^{\mathrm{SM}}$ |
| $S_3 = (\bar{\mathbf{3}}, \mathbf{3})_{1/3}$ | ✓ | ✗ | $R_{D^{(*)}}$ vs. $B \to K^{(*)} \nu \bar{\nu}$ |
| $U_1 = (\mathbf{3}, \mathbf{1})_{2/3}$ | ✓ | ✓ | |
| $U_3 = (\mathbf{3}, \mathbf{3})_{2/3}$ | ✓ | ✗ | $R_{D^{(*)}}$ vs. $B \to K^{(*)} \nu \bar{\nu}$ |

whereas they do not contribute at the tree level to four-quark or four-lepton contact interactions, which so far do not exhibit deviations from the SM. A detailed comparison of how well each mediator addresses each of the two anomalies has been presented in Ref. [368]. A concise summary of this study is reported in Table 4.4. As can be seen, there is a single field able to account for both anomalies, namely the $U_1$ leptoquark.

The failure of a given mediator in describing both anomalies can be overcome considering models with two or more mediators. Two notable examples of this type are: i) the $S_1$–$S_3$ solution proposed by Marzocca [369], in the framework of a composite model where these two states arise as pseudo Nambu-Goldstone bosons of a new strongly coupled sector; ii) the $R_2$–$S_3$ solution proposed by Becirevic *et al.*, in the framework of a $SU(5)$ unified model [350]. Note also that if the charged-current anomalies would result into a mere statistical fluctuation, the $S_3$ scalar leptoquark would be an excellent candidate to explain the netural-current anomalies alone. However, if both anomalies were confirmed as clear signals of physics beyond the SM, it is fair to say that the $U_1$ solution emerges for simplicity and, especially, reduced tuning, compared to all the other options, at least at the level of simplified models.

The key point that speaks in favor of the $U_1$ solution compared to any other option (such as the two-mediators frameworks mentioned above), is the absence of large deviations from the SM in $B \to K^{(*)} \nu \bar{\nu}$ decays. Being forbidden at the tree level within the SM, $B \to K^{(*)} \nu \bar{\nu}$ decays set a stringent constraint on the $b_L \to s_L \nu_\tau \bar{\nu}_\tau$ amplitude, which involves a $3 \rightarrow 2$ transition in the quark sector and left-handed leptons of the third generation. The fields involved are the same entering in the effective operator $\mathcal{O}_{LL}^{23\tau\tau}$ in Eq. (4.26), but with a different contraction of the $SU(2)_L$ indices (corresponding to the combination $Q_{lq}^{(1)} - Q_{lq}^{(3)}$ in the SMEFT Warsaw basis). The need of a large coefficient for $\mathcal{O}_{LL}^{23\tau\tau}$, in order to address the $b \rightarrow c\tau\bar{\nu}$ anomalies, and the requirement of a small coefficient for the effective operator contributing to the $b_L \rightarrow s_L \nu_\tau \bar{\nu}_\tau$ amplitude is a non-trivial constraint. The $U_1$ field, which couples down-type quarks to charged leptons, and up-type quarks to neutrinos, provides a tree-level contributions to $b \rightarrow s\ell^+\ell^-$ and $b \rightarrow c\tau\bar{\nu}$ amplitudes and does not generate a tree-level contribution to $b \rightarrow s\nu\bar{\nu}$. In all the other cases, this result is achieved only via a significant tuning of the model parameters.

Beside this phenomenological aspect, the $U_1$ case has a strong advantage from the model building point of view, finding a convincing motivation in models based on the $SU(4)$ group unifying quark and leptons proposed by Pati and Salam in 1974 [360]. Given these arguments, in section 4.6 we investigate in more detail the consequences of the assumption that both anomalies are the result of the exchange of a single $U_1$ field.

### 4.5.2 *Colorless mediators*

The possibility of describing both set of anomalies in terms of colorless mediators, in particular an $SU(2)_L$ triplet ($W'$ and $Z'$) [341] or a combination of $SU(2)_L$ triplets and singles, which appeared to be feasible a few years ago, is nowadays basically excluded by the detailed analysis of electroweak precision observables and the strong bounds from high-energy physics [359].

However, colorless mediators remain valuable candidates for separate (independent) explanations of the two anomalies. Concerning charged currents, an interesting option is that of a heavy $W'$ field coupled to right-handed fermions in conjunction with a light right-handed sterile neutrino (see Ref. [370–373]). In this setup the observed $b \rightarrow c\tau\nu$ rates are enhanced not because of a change in the leading left-handed amplitude, but because of an additional (non-interfering) channel involving the sterile neutrino. As shown in Ref. [370], despite being tightly constrained, especially by direct searches, this option is still open, and the presence of a sterile neutrino might link it to the explanation of some of the cosmological problems of the SM. On the phenomenological side, this solution of the charged-current anomalies could be detected by precision studies of the polarization asymmetries in $B \to D^* \tau \nu$, which are expected to deviate from the predictions shown in Fig. 4.5.

Concerning the charged-current anomalies, a heavy $Z'$ boson with lepton-conserving (non-universal) couplings to lepton is a particularly interesting option. Heavy $Z'$ bosons appear in motivated extensions of the SM featuring extra $U(1)$ gauge symmetries (see in particular [285, 308–318]). The absence of flavor-violating couplings to leptons automatically avoids the tight bounds from Lepton Flavor Violating (LFV) processes; however, viable models require a somewhat large ratio of the $Z'$ couplings to leptons vs. quarks in order to avoid the stringent bounds from $B_s$ mixing (which is generated at the tree level) [374].

## 4.6    A closer look to the $U_1$ leptoquark

As mentioned above, the simplified model based on the $U_1$ field is the most interesting option for a combined description of the two sets of anomalies. Introducing a new heavy vector boson necessarily requires the introduction of an additional new-physics sector, describing the mechanism through which the vector acquire its mass. We postpone a detailed discussion about this and other model-building aspects to the next section. Here we limit to consider a simplified model where a massive $U_1$ field is responsible for the new flavor-changing LFU-violating interactions.

The most general Lagrangian for a $U_1$ vector leptoquark coupled to SM particles is

$$
\mathcal{L}_U = -\frac{1}{2} U_{\mu\nu}^\dagger U^{\mu\nu} + M_U^2 U_\mu^\dagger U^\mu - i g_s (1 - \kappa_c) U_\mu^\dagger T^a U_\nu G^{\mu\nu,a}
$$

$$
- \frac{2i}{3} g_Y (1 - \kappa_Y) U_\mu^\dagger U_\nu B^{\mu\nu} + \frac{g_U}{\sqrt{2}} (U^\mu J_\mu^U + \text{h.c.}), \qquad (4.34)
$$

where $U_{\mu\nu} = D_\mu U_\nu - D_\nu U_\mu$, with $D_\mu = \partial_\mu - i g_s G_\mu^a T^a - i \frac{2}{3} g_Y B_\mu$. Here $G_\mu^a$ ($a = 1, \ldots, 8$) and $B_\mu$ denote the $SU(3)_c$ and $U(1)_Y$ gauge bosons, $g_s$ and $g_Y$ are the corresponding gauge couplings, and $T^a$ are the generators of $SU(3)_c$. In models in which the vector leptoquark has a gauge origin, $\kappa_c = \kappa_Y = 0$, while this is not necessarily the case in models in which the $U_1$ arises as a bound state from a strongly-coupled sector. The interaction of the $U_1$ with the SM fermions involves the currents

$$
J_\mu^U = \beta_L^{i\alpha} (\bar{q}_L^i \gamma_\mu \ell_L^\alpha) + \beta_R^{i\alpha} (\bar{d}_R^i \gamma_\mu e_R^\alpha), \qquad (4.35)
$$

where the couplings $\beta_L$ and $\beta_R$ are complex $3 \times 3$ matrices in flavor space. Following our hypothesis on the flavor structure of the theory, we can write

$$\beta_L = \begin{pmatrix} 0 & 0 & \beta_L^{d\tau} \\ 0 & \beta_L^{s\mu} & \beta_L^{s\tau} \\ 0 & \beta_L^{b\mu} & 1 \end{pmatrix}, \quad \beta_R = \begin{pmatrix} 0 & 0 & 0 \\ 0 & 0 & 0 \\ 0 & 0 & \beta_R^{b\tau} \end{pmatrix}, \tag{4.36}$$

with $|\beta_L^{d\tau,s\mu}| \ll |\beta_L^{s\tau,b\mu}| \ll 1$ and $\beta_L^{b\tau} = \mathcal{O}(1)$. The normalization of $g_U$ is chosen such that $\beta_L^{b\tau} = 1$. The null entries in Eq. (4.36) should be understood as small terms (according to our scaling, they are expected to be all below $10^{-2}$) which have a negligible impact on the observables we analyze (i.e. mainly $3 \to 2$ flavor transitions in the quark and lepton sectors). It is worth stressing that this structure is a direct consequence of the hypothesis of a $U(2)^5$ flavor symmetry with sizable breaking only along the $U(2)_q$ direction.[4] Under the stronger assumption of a single spurion transforming as doublet of $U(2)_q \in U(2)^5$, we further expect $\beta_L^{d\tau}/\beta_L^{s\tau} = V_{td}^*/V_{ts}^*$ (see appendix B).

By integrating out the vector leptoquark at tree level, we obtain the following matching conditions for the effective operators introduced in section 4.4.2:

$$C_{LL}^{ij\alpha\beta} = C_U \beta_L^{i\alpha} (\beta_L^{j\beta})^*, \quad C_{LR}^{ij\alpha\beta} = C_U \beta_L^{i\alpha} (\beta_R^{j\beta})^*, \quad C_{RR}^{ij\alpha\beta} = C_U \beta_R^{i\alpha} (\beta_R^{j\beta})^*, \tag{4.37}$$

where $C_U \equiv g_U^2 v^2 / (4 M_U^2)$. The coupling $C_U$ determines the normalization of the $U_1$-induced left-handed semileptonic interactions, for third-generation fermions, relative to the Fermi coupling $G_F$.

### 4.6.1 *Fit of low-energy observables*

Using the matching conditions in Eq. (4.37) we can proceed fitting all available experimental data on semileptonic processes, as well as other low-energy observables, and determine the values of the $\beta_{L(R)}^{i\alpha}$ and the overall coupling $C_U$. As already mentioned in section 4.4.3, beside the anomalies, the most important low-energy constraints are the precise tests of universality in leptonic $\tau$ decays [375] and the bounds from $\Delta F = 2$ amplitudes (see section 2.4). In both these cases we must proceed as in

---

[4]As far as right-handed mixing is concerned, this symmetry hypothesis alone implies that the natural size of the largest off-diagonal entries is $\beta_R^{s\tau} \sim (m_s/m_b) \beta_L^{s\tau}$ and $\beta_R^{b\mu} \sim (m_\mu/m_\tau) \beta_L^{s\tau}$, well below the corresponding left-handed entries.

the EFT analysis, introducing an appropriate cut-off to regularise the UV-divergent contributions generated by the $U_1$ exchange beyond the tree-level. Additional important low-energy constraints are derived by FCNC $B$ decays into $\tau^+\tau^-$ pairs [376, 377], which are expected to be largely enhanced compared to the SM expectations, LFV $B$ decays of the type $b \to s\tau^\pm\mu^\mp$ [378, 379], and LFV $\tau^-$ decays such as $\tau \to \mu\gamma$ [375] and $\tau \to \mu\phi$ [380].

The parameter $\beta_R^{b\tau}$ controls the amount of right-handed currents in $b \to c\tau\bar{\nu}$ transitions. As discussed in section 4.3.2, present data are not very conclusive on the presence of right-handed currents, correspondingly we cannot set stringent experimental constraints on $\beta_R^{b\tau}$. For this reason, it is convenient to consider the following two reference cases: $\beta_R^{b\tau} = 0$ and $\beta_R^{b\tau} = -1$. As we shall discuss in the next section, in models with third-family Pati-Salam unification, and in absence of a mixing of the SM fermions with exotic fermions, one expects $|\beta_R^{b\tau}| = 1$. On the other hand, a value $|\beta_R^{b\tau}| \ll 1$ can be obtained, for instance, in models where the $U_1$ is a composite state. The condition $|\beta_R^{b\tau}| = 1$ does not fix the phase of $\beta_R^{b\tau}$. In the large-$\beta_R^{b\tau}$ scenario we set $\beta_R^{b\tau} = -1$ in order to maximize the constructive interference of left-handed and right-handed contributions in the charged-current anomalies ($\beta_R^{b\tau} = -1$ corresponds to the line $C_{LR}^c = -C_{LL}^c$ in Fig. 4.4). For a similar reason, the coupling $\beta_L^{s\tau}$ is assumed to be real.

In Fig. 4.8 (left) we report the values of the $U_1$ effective couplings obtained for $\beta_R^{b\tau} = 0$ and setting $\beta_L^{d\tau}/\beta_L^{s\tau} = V_{td}^*/V_{ts}^*$. In such case one obtains a $\Delta\chi^2 \approx 59$ between SM and $U_1$ hypotheses, where the NP case has only four additional degrees of freedom ($N_{\rm dof}$). The $\Delta\chi^2/N_{\rm dof}$ is slightly lower in the case $\beta_R^{b\tau} = -1$, or if we enhance the number of degrees of freedom letting $\beta_L^{d\tau}/\beta_L^{s\tau}$ to vary, both in magnitude and in phase. However, in all cases one obtains a significance well above $5\sigma$ for the $U_1$ simplified model compared to the SM case.

A clear illustration of this statement is presented in Fig. 4.9, where we show the $1\sigma$ and $2\sigma$ preferred regions for the LFU-violating observables

$$\delta R_{K^{(*)}} = \frac{R_{K^{(*)}} - R_{K^{(*)}}^{\rm SM}}{R_{K^{(*)}}^{\rm SM}}, \quad \delta R_{D^{(*)}} = \frac{R_{D^{(*)}} - R_{D^{(*)}}^{\rm SM}}{R_{D^{(*)}}^{\rm SM}}, \tag{4.38}$$

obtained for the two reference options $\beta_R^{b\tau} = 0$ and $\beta_R^{b\tau} = -1$. As can be seen, the overall agreement with data is very good in both cases. The $\beta_R^{b\tau} = -1$ case is only slightly disfavored because of the lower prediction of $R_{D^*}$ compared to data.

| Parameter | best fit | $1\sigma$ |
|-----------|----------|-----------|
| $C_U$ | 0.010 | $[0.007, 0.017]$ |
| $\beta_L^{b\mu}$ | $-0.15$ | $[-0.26, -0.02]$ |
| $\beta_L^{s\tau}$ | 0.19 | $[0.10, 0.25]$ |
| $\beta_L^{s\mu}$ | 0.014 | $[0.004, 0.14]$ |

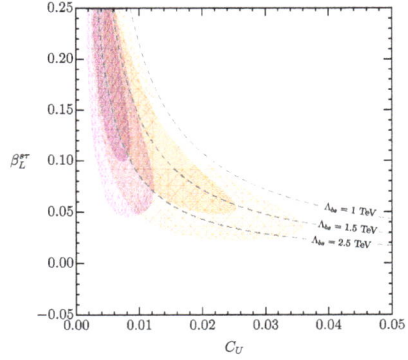

Fig. 4.8: Low-energy constraints on the $U_1$ effective couplings. *Left*: preferred values of the parameters extracted by a global fit to low-energy data [300], setting $\beta_R^{b\tau} = 0$ and $\beta_L^{d\tau}/\beta_L^{s\tau} = V_{td}^*/V_{ts}^*$. *Right*: Correlation between $C_U$ and $\beta_L^{s\tau}$ for $\beta_R^{b\tau} = 0$ (orange) and $\beta_R^{b\tau} = -1$ (purple). The darker and lighter regions denote $\Delta\chi^2 \leq 2.30$ ($1\sigma$) and $\Delta\chi^2 \leq 6.18$ ($2\sigma$), respectively. The dotted lines illustrate the impact of $B_s$ mixing.

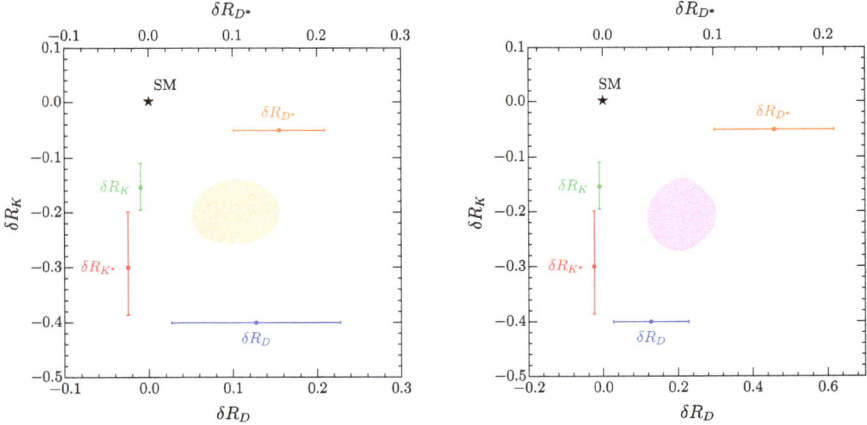

Fig. 4.9: Preferred $1\sigma$ and $2\sigma$ regions for the ratios $\delta R_{D^{(*)}}$ (upper horizontal scale), $\delta R_D$ (lower horizontal scale), and $\delta R_{K^*} \approx \delta R_K$ (vertical scale), resulting from the low-energy fit of the $U_1$ model for $\beta_R^{b\tau} = 0$ (orange) and $\beta_R^{b\tau} = -1$ (purple). The colored error bars show the current experimental measurements at $1\sigma$ (figure from Ref. [300]).

The key results of the low-energy fit can be summarised as follows:

- The fitted values of the $U_1$ couplings are perfectly consistent with the flavor structure discussed in the previous sections: $|\beta_L^{s\tau}| \sim 0.1$, $|\beta_L^{b\mu}| \sim 0.1$, $|\beta_L^{s\mu}| \sim 0.01$.

- The value of $C_U$ is in the $10^{-2}$ range, and there is a strong correlation between $C_U$ and $\beta_L^{s\tau}$, illustrated in Fig. 4.8 (right). This is a direct consequence of the dependence of $\delta R_{D^{(*)}}$ from $C_{LL}^{23\tau\tau} = C_U \times \beta_L^{s\tau}$.
- The minimal $U(2)^5$-breaking relation

$$\beta_L^{d\tau}/\beta_L^{s\tau} = V_{td}^*/V_{ts}^*, \tag{4.39}$$

is well supported by data; however, $\mathcal{O}(1)$ deviations in both magnitude and phase are possible.
- The most severe low-energy constraint is posed by $B_s$ mixing. As shown in Fig. 4.8 (right) this can be satisfied only if the effective UV cut-off on the $B_s$ mixing amplitude ($\Lambda_{bs}$) does not exceed $\sim 1$ TeV. As we will mention in section 4.7, this can be achieved with vector-like leptons with mass $\sim \Lambda_{bs}$.

### 4.6.2 *Leptoquark signals at high energies*

Having determined the $U_1$ couplings from low-energy data, we can analyse the impact of this new field in high-energy observables. The partonic diagrams corresponding to the most relevant high-energy signatures of the $U_1$ at the LHC, i.e. pair production, single production, and $t$-channel exchange, are shown Fig. 4.10.

The leptoquark pair-production is dominated by QCD dynamics [351, 353, 381, 382] [Fig. 4.10 (a)] and thus it is largely independent of the leptoquark couplings to fermions. A residual model dependence is introduced by the possible non-minimal couplings to gluons, parameterized in the Lagrangian in Eq. (4.34) by the parameter $\kappa_c$. In models where the vector leptoquark has a gauge origin, this non-minimal coupling is absent ($\kappa_c = 0$), allowing for robust theory predictions for the pair-production cross section. Given the flavor structure emerging from low-energy data, the dominant decay channels are those involving pairs of third-generation

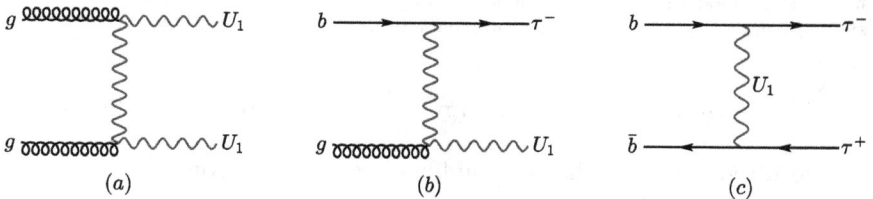

Fig. 4.10: Representative Feynman diagrams for vector leptoquark pair production (a), single-leptoquark production (b), and $t$-channel Drell-Yan production (c).

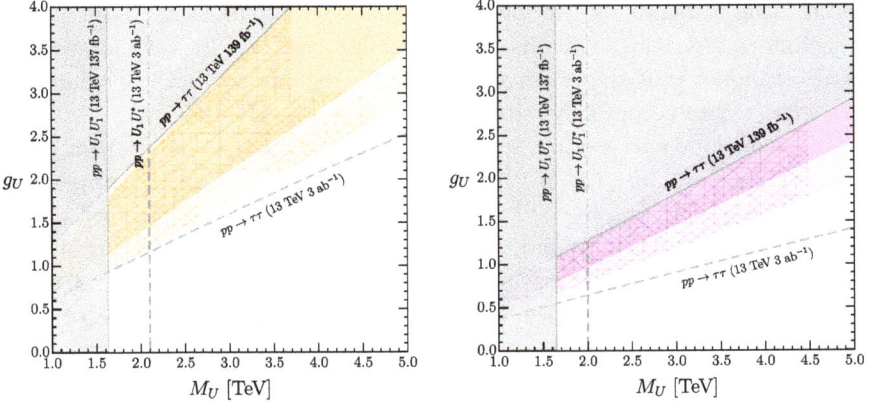

Fig. 4.11: LHC constraints for the $U_1$ vector leptoquark for the benchmark scenarios with $\beta_R^{b\tau} = 0$ (left) and $\beta_R^{b\tau} = -1$ (right) at the end of 2021. The colored bands indicate the $1\sigma$ and $2\sigma$ preferred regions from the fit to low-energy data illustrated in Fig. 4.8–4.9 (figure from Ref. [300]).

fermions, namely $U_1 \to b\tau^+$ and $U_1 \to t\bar{\nu}$. In Fig. 4.11 we show the present bounds derived from pair production via the channel $pp \to U_1^* U_1 \to b\tau t\nu$ analysed by CMS [383], together with the $1\sigma$ and $2\sigma$ regions obtained from the low-energy fit. In the same figure the projected limits for the high-luminosity phase of the LHC (assuming $3\,\mathrm{ab}^{-1}$ of integrated luminosity and no signal) are also shown. As can be seen, these searches offer only a relatively small coverage of the parameter space favored by the low-energy fit.

The most interesting collider constraint is obtained by searching for modifications of the high-$p_T$ tail in the dilepton invariant mass distribution in the Drell-Yan process $pp \to \tau^+\tau^- + X$ induced by $t$-channel $U_1$ exchange (Fig. 4.10 c), which has been pioneered in Ref. [357] (see also [352, 359, 382, 384]). The dominant production mechanism for this channel is via a $b\bar{b}$ initial state. Contributions from $b\bar{s}$- and $s\bar{s}$-initiated processes are subdominant due to the underlying flavor structure of the leptoquark couplings. In absence of a dedicated study of this channel by ATLAS and CMS, interesting limits can be obtained by recasting existing searches of exotic Higgs bosons decaying into $\tau^+\tau^-$, as described in [382]. Using the 2020 ATLAS analysis in [385] with $139\,\mathrm{fb}^{-1}$ of 13 TeV data leads to the bounds shown in Fig. 4.11, from Ref. [300]. As can be seen, the high-$p_T$ lepton tails provide significant constraints on the parameter space preferred by the low-energy fit, especially for $\beta_R^{b\tau} = -1$, where the limit is stronger due to the larger production cross section. On the other hand, a large region of the parameter space still remains viable for both benchmark scenarios, which is a highly non trivial result. Very interesting is also the fact that the preferred $1\sigma$ and $2\sigma$ regions, for both benchmarks, are expected to completely within the reach of the HL-LHC.

To conclude this section, it is worth mentioning that limits from $pp \to \mu\tau$ [382, 386] and $pp \to \mu^+\mu^-$ [358] do not provide, at present, competitive bounds compared to $pp \to \tau^+\tau^-$ because of the flavor suppression of the light-lepton couplings. Similarly, limits derived from the $t$-channel exchange from $b$ and $c$ initial partons, namely via $pp \to \tau\bar{\nu}$ [387], are found to be weaker compared to those set by $pp \to \tau^+\tau^-$ due to the smallness of $|\beta_L^{s\tau}|$ and $|V_{cb}|$. Finally, also the bounds from the single resonant production via quark-lepton fusion ($pp \to \bar{b}\tau$), which are potentially interesting in specific regions of the parameter space [388, 389], turn out to be weaker compared to those set by $pp \to \tau^+\tau^-$ in the region of interest to explain both anomalies. However, all these subleading channels might play a relevant role in the future, in the event of a discovery.

### 4.6.3   *Implications for low-energy observables*

The $U_1$ simplified model implies clear non-standard predictions for a series of low-energy observables different from the present anomalies. While four-quark or four-lepton processes are largely dependent on the UV completion of the model, selected semileptonic and LFV decays can be predicted with high accuracy. The two most interesting effects are:

- large rates, within experimental reach, for LFV processes in the $\tau \leftrightarrow \mu$ sector, such as $B^+ \to K^+\tau^+\mu^-$ and $B_{s,d} \to \tau^-\mu^+$, as well as $\tau \to \mu\gamma$ and $\tau \to \mu\phi$;
- large enhancements over the SM predictions for $b \to s$ transitions involving a $\tau^+\tau^-$ pair, such as $B_{s,d} \to \tau^+\tau^-$ and $B \to K^{(*)}\tau^+\tau^-$.

The predictions for some of these processes resulting from the low-energy fit of the anomalies are shown in Fig. 4.12. With the exception of $\tau \to \mu\gamma$, all these observables are dominated by the tree-level contribution from the $U_1$ exchange. This partially true also for $\tau \to \mu\gamma$ that, despite being generated at the loop level, is largely insensitive to the UV completion of the model in the large $|\beta_R^{b\tau}|$ regime (which is the only case where it can be reach large values close to present bounds [300, 368, 390]). In principle, also LFV processes in the $\mu \leftrightarrow e$ sector are generated at the tree level and could be close to present bounds; however, in this case is more difficult to make

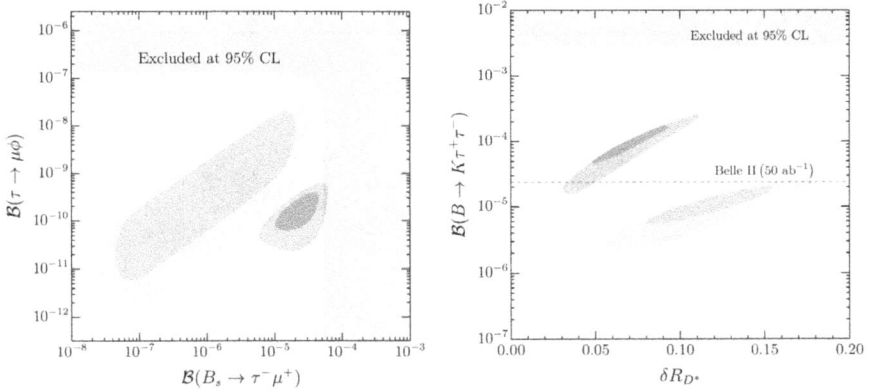

Fig. 4.12: Preferred $1\sigma$ and $2\sigma$ regions for the branching fractions of the LFV processes $B \to \tau\phi$ versus $B_s^0 \to \tau^-\mu^+$ (left) and $\mathcal{B}(B^+ \to K^+\tau^+\mu^-)$ versus $\delta R_{D*}$ (right) expected within the $U_1$ simplified model for the benchmark scenarios with $\beta_R^{b\tau} = 0$ (orange/light gray) and $\beta_R^{b\tau} = -1$ (purple/dark gray). The region denoted 'Excluded at 95% CL' indicate the experimental limits in 2021 (from Ref. [300]).

precise predictions due to lack of information on the $U_1$ couplings to first-generation leptons.

All these observables represent unambiguous signals of physics beyond the SM which would provide a clear confirmation of this model. If $|\beta_R^{b\tau}|$ is large, the predictions for $B_s \to \tau^+\tau^-$, $B \to K\tau^+\tau^-$, $B_s \to \tau^-\mu^+$, and $\tau \to \mu\gamma$ are very close to present bounds. This is because of the chiral-enhancement of the corresponding amplitudes. We recall that the two benchmarks we have chosen for $|\beta_R^{b\tau}|$ are representative of two extreme options, and any intermediate value can be considered as a viable option.

Beside these large deviations from the SM predictions, a series of interesting clean predictions can be made also on other processes where the NP effect is only a small correction over the SM expectation. For instance the LFU ratio $R_\pi = \mathcal{B}(B^+ \to \pi^+\mu^+\mu^-)/\mathcal{B}(B^+ \to \pi^+e^+e^-)$ is an ideal probe of the coupling $\beta_L^{d\tau}$: if Eq. (4.39) is satisfied, then the model predict $R_\pi \approx R_K$ [391]. The analog in the charged-current sector are $\mathcal{B}(\bar{B}_u \to \tau\bar{\nu})$ and $\mathcal{B}(\bar{B} \to \pi\tau\bar{\nu})$, whose predictions are shown in Fig. 4.13 assuming the relation in Eq. (4.39). In this case a further source of uncertainty is represented by the impact of right-handed currents: this is why we show these observables as a function of $\delta R_D - \delta R_{D*}$, which is also controlled by right-handed currents.

The only class of semileptonic decays for which is difficult to draw precise predictions in the absence of a clear UV completion for the $U_1$

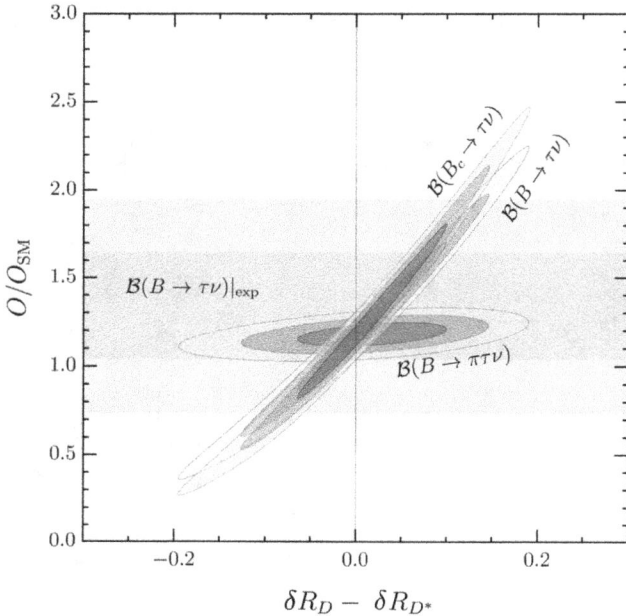

Fig. 4.13: Predictions for $B(\bar{B}_c \to \tau\bar{\nu})$, $B(\bar{B}_u \to \tau\bar{\nu})$, and $\mathcal{B}(\bar{B} \to \pi\tau\bar{\nu})$, all normalized to the corresponding SM expectations, as a function of $\delta R_D - \delta R_{D^*}$. The horizontal gray bands denote the experimental value of $B(\bar{B}_u \to \tau\bar{\nu})$ at $1\sigma$ and $2\sigma$ (figure from Ref. [323]).

are $b \to s\nu\bar{\nu}$ transitions. As already discussed, the $U_1$ cannot mediate a tree-level contribution to the $b_L \to s_L \nu_\tau \bar{\nu}_\tau$ amplitude; however, sizeable contributions are expected beyond the tree level. If the $U_1$ as a gauge origin, as in the class of models discussed in section 4.7.1, it is possible to establish a clear correlation between $\mathcal{B}(B \to K^{(*)}\nu\bar{\nu})$ and $\delta R_{D^*}$. Interestingly enough, the unambiguous expectation is that of an enhancement of $\mathcal{B}(B \to K^{(*)}\nu\bar{\nu})$, compared to its SM expectation, which cloud reach up to about 50% [365].

## 4.7    Ultraviolet-complete models

In the previous two sections we discussed some of the possible heavy mediators responsible for the two sets of anomalies, focusing in particular on the $U_1$ case, which appears to be a particularly interesting option for a combined explanation. Identifying the mediator, or the set of mediators, is only the first step toward the identification of a consistent UV extension of the SM. The latter is a more ambitious and, at the same time, much more difficult objective. Still, it is worth to sketch some of the possibilities which have been proposed in the last few years starting from simplified models addressing the flavor anomalies.

The two main points to address in order to identify a consistent extension of the SM, starting from the flavor anomalies, are:

I. Dynamical origin of the mediators. Which is the minimal consistent set of new fields to be introduced at a given scale, and which is the rational behind their masses? We address this point in section 4.7.1, devoted to the $U_1$ case, and section 4.7.2, where we briefly comment on alternative options.

II. Origin of the flavor structure of the model. This is clearly a very important point for models addressing the flavor anomalies, and is somehow independent from the nature of mediators. As we discuss in section 4.7.3, this point could be connected to a possible explanation of the SM flavor hierarchies.

### 4.7.1 *The $U_1$ field and quark-lepton unification*

Focusing the attention to the $U_1$ case as leading mediator for the anomalies, the open questions we need to address from a model-building perspective are: which is the origin of its mass (being a massive spin-one particle)? Is it consistent to consider only the $U_1$ as new state at the TeV scale?

**The embedding of the $U_1$ in $SU(4)$**

As anticipated, the $U_1$ field naturally points to the $SU(4)$ symmetry of the Pati-Salam (PS) gauge group, unifying quark and lepton quantum numbers [360]. The complete PS gauge group is $SU(4) \times SU(2)_L \times SU(2)_R$. Under this group, a complete family of quarks and leptons, together with a right-handed neutrino, is unified into two fundamental fermions $\psi_{L,R}$,

$$\psi_L = \begin{pmatrix} q_L^\alpha \\ q_L^\beta \\ q_L^\gamma \\ \ell_L \end{pmatrix}, \quad \psi_R = \begin{pmatrix} q_R^\alpha \\ q_R^\beta \\ q_R^\gamma \\ \ell_R \end{pmatrix}, \tag{4.40}$$

transforming as $(\mathbf{4}, \mathbf{2}, \mathbf{1})$ and $(\mathbf{4}, \mathbf{1}, \mathbf{2})$, respectively. Here the greek labels denote ordinary color, and right-handed fields are organised into doublets of $SU(2)_R$: $q_R^T = (u_R, d_R)$ and $\ell_R^T = (\nu_R, e_R)$. In this context lepton number emerges as the *fourth color*, as originally stated by Pati and Salam [360].

One of the most appealing features of the PS group is the explanation of the quantized ratio between the electric charges of quarks and leptons. Within the SM, this is somehow imposed via the *ad hoc* hypercharge assignments in Table 1.1. Within the PS group, the SM hypercharge is nothing but the combination $Y = T_R^3 + \frac{1}{2}T_{B-L}$, where $T_R^3$ and $T_{B-L}$ are diagonal of generators of $SU(2)_R$ and $SU(4)$, respectively.[5] The electric charge of any fermion is then automatically quantized (in units of the electron charge) and is expressed by the simple formula

$$Q = T_L^3 + T_R^3 + \frac{1}{2}T_{B-L}. \tag{4.41}$$

The breaking PS → SM can be decomposed into three separate steps:

i. $SU(2)_R \to U(1)_{T_R^3} \qquad\qquad \longrightarrow W_R^\pm$

ii. $SU(4) \to SU(3)_c \times U(1)_{B-L} \longrightarrow U_1 \qquad (4.42)$

iii. $U(1)_{T_R^3} \times U(1)_{B-L} \to U(1)_Y \longrightarrow Z'$

where we have shown the gauge fields acquiring a mass, assuming a spontaneous symmetry breaking of the gauge symmetry. As can be seen, a massive $U_1$ field is generated by the breaking of $SU(4)$ into ordinary color. The breaking of $SU(2)_R$ (step i.) can occur at high scales and it does not necessarily imply additional vectors states close in mass to the $U_1$. On the other hand, the $Z'$ boson (step iii.) is necessarily close in mass to the $U_1$, or lighter, since it also arises from the breaking of generators of the $SU(4)$ group commuting with $SU(3)_c$.

From the above discussion, we conclude that a massive $U_1$ is necessarily accompanied by a $Z'$ close in mass, coupled to $(B - L)$ quantum numbers, at least if these massive vectors arises from the breaking of the $SU(4)$ gauge group. As discussed in [382], this conclusion is more general: it holds also if the $U_1$ field is a spin-one resonance within a composite sector addressing the electroweak hierarchy problem, where $SU(4)$ is only a global symmetry, as proposed for instance in Ref. [344, 392, 393].

### From Pati Salam to 4321

The $U_1$ field of the PS model has the correct quantum numbers (under the SM gauge group) to explain the anomalies, but not the right flavor structure.

---

[5]$T_{B-L}$ is the diagonal generator of $SU(4)$ commuting with color, whose action on quarks and lepton reads $T_{B-L}\, q = \frac{1}{3}q$ and $T_{B-L}\, \ell = -\ell$.

The PS model is indeed a flavor-blind model where the three generations appear as three identical copies, but for Yukawa couplings, exactly as in the SM case. On the other hand, an explanation of the flavor anomalies require a $U_1$ field coupled predominantly to the third generation: a $U_1$ field with large couplings to the light families is necessarily pushed to high scales by the severe bounds on processes such as $K_L \to \mu e$, $K \to \mu \mu$, and $\mu N \to Ne$ (see e.g. [394, 395]) and it cannot provide a sizable contribution to the $B$ anomalies.

If the $U_1$ is a massive gauge boson, a non-trivial flavor structure can be achieved in two ways: extending the fermion content or extending the gauge group. The first option, pursued for instance in Ref. [396], requires a severe tuning in the parameter space. The second option, discussed in several recent works [364, 395, 397–403], seems to be more appealing from the model-building point of view.

Independently of the $b$-physics anomalies, in Ref. [351, 404] it has been noted that the color group could appear as diagonal subgroup of a larger $SU(3+N) \times SU(3)$ local symmetry valid at high energies. Di Luzio $al.$ [397] realised that this set up was ideal to address the problem of the $U_1$ flavor structure in the context of the $B$ anomalies via an extension of the PS symmetry. More precisely, the $SU(4)$ group of PS is extended to $SU(4) \times SU(3)'$, and the breaking chain

$$SU(4) \times SU(3)' \to SU(3)_c \times U(1)_{B-L} \quad \longrightarrow \quad U_1, G' \qquad (4.43)$$

replaces the second step in Eq. (4.42). This extension has the advantage of separating the LQ coupling from the ordinary strong coupling. Most important, this enlargement of the group allow us to achieve quark-lepton unification in a flavor non-universal manner [398, 401, 403, 405], charging only the third family under $SU(4)$. This way the approximate $U(2)^5$ flavor symmetry emerges as accidental flavor symmetry of the gauge sector, given the first two generations have different transformation properties with respect to the third one:

$$\left. \begin{array}{ll} 1^{st}, 2^{nd} \text{ generation}: & SU(3)' \\ 3^{3d} \text{ generation}: & SU(4) \end{array} \right\} \times SU(2)_L \times U(1)',$$

$$U(1)' = \begin{cases} U(1)_Y \\ U(1)_{T_R^3} \end{cases} \qquad (4.44)$$

The important price to pay for this class of models, known as 4321 models, is the appearance of a new TeV-scale vector with the quantum numbers of a gluon $(G')$, coupled predominantly to third-generation quarks. Once more, this conclusion hold also in the context of composite models where $SU(4) \times SU(3)'$ is only a global symmetry [393].

It is interesting to note the analogy of the $SU(4) \times SU(3)'$ group for the strong-LQ sector, with the $SU(2)_L \times U(1)_Y$ group for the electroweak sector. In the latter case we achieve a unified description of weak and electromagnetic interactions, whereas in the former case we unify strong and (superweak) LFU-violating interactions. In both cases the unbroken symmetries surviving at low energies are the corresponding diagonal subgroups.

The enlargement of the gauge group leads to a desired flavor structure for the leading $U_1$ couplings, but it does not prevent an enlargement of the fermion sector of the theory. In all realistic models one or more families of vector-like fermions, with masses in the TeV domain, are needed to describe subleading flavor-mixing terms (including the subleading entires in the SM Yukawa couplings). The TeV spectrum of these UV consistent models is therefore much richer than what could be deduced by looking only at the corresponding simplified model. However, it is worth stressing that, at least for the gauge models, these UV-complete constructions allow us to systematically evaluate low- and high-energy amplitudes beyond the tree-level in terms of a limited number of free parameters [365, 406, 407]. For instance, the detailed analysis of $B_s$ mixing implies the need of vector-like partners for the $\tau$ lepton with mass not exceeding $1.5\,\text{TeV}$ [300].

### 4.7.2 *Other options: scalar leptoquarks and $Z'$ bosons*

As discussed in section 4.5 the other classes of interesting mediators, especially if only the $b \to s\ell^+\ell^-$ anomalies are considered, are scalar leptoquarks and $Z'$ bosons.

Concerning scalar LQs, a very interesting possibility is to consider these states as the pseudo-Goldstone bosons of a new strongly interacting sector, with a confining scale in the 10 TeV regime, possibly generating also the Higgs field as pseudo-Goldstone boson. This way the $B$-anomalies would naturally be linked to the stabilization of the Higgs sector and the solution of the electroweak hierarchy problem. The pseudo-Goldstone nature of the scalar LQs, which would resemble the pions in QCD, would also explain why these states are the lightest exotic states above the SM spectrum. Proposals of this type have been discussed for instance in Ref. [346, 369]. The problem in this context is that is impossible to conceive a setup (or better a coset corresponding to the breaking of a global symmetry in the strong sector) where only the Higgs and the LQs needed to explain the anomalies emerge as light pseudo-Goldstone bosons: other light TeV states are naturally present, and these might be problematic for direct and indirect searches [369]. However, there is no doubt that pseudo-Goldstone LQs is a very interesting option that is still open and worth to be pursued.

Fundamental scalar LQs could also arise by a variety of grand-unified models. Interesting proposal to host scalar LQs able to address at least the $b \to s\ell^+\ell^-$ anomalies have been made for instance in Ref. [408, 409], in the context of PS-like models, and in Ref. [350], in the framework of a $SU(5)$ unified model. What these proposal misses is a convincing justification for the non-trivial structure for the couplings of these scalar fields. But this is not a pressing argument, especially when focusing only one set of anomalies.

Concerning $Z'$ bosons addressing the $b \to s\ell^+\ell^-$ anomalies, an interesting option is to link them to possible anomaly-free flavor-dependent $U(1)$ gauge symmetries, such as $L_\tau$-$L_\mu$ or other similar combinations [310–313, 315, 317, 318]. This possibility is less connected to the known problems of the SM: mass and couplings of the $Z'$ needed to explain the anomalies are somewhat more *ad hoc*. On the other hand, as pointed out recently in Ref. [410], this type of set up could point to a link between the $B$-anomalies and the discrepancy between data and SM predictions for the anomalous magnetic moment of the muon [411].

The connection between $b$-anomalies and $(g - 2)_\mu$, which has been investigated also in Ref. [412, 413], is far from obvious: the $(g-2)_\mu$ anomaly, if confirmed, could be addresses by a different NP sector (e.g. some new *light* state). But it is certainly worth to be explored further.

### 4.7.3   A multi-scale picture at the origin of the flavor hierarchies

The most important observation about the flavor structure of the new interactions is summarised by Table 4.3: the hierarchical values for coefficients of the three operators in this table, together with the well-known observation that physics beyond the SM in the TeV domain must mimic the favor structure of the SM to be consistent with data [155], seem to point to a common origin of flavor anomalies and SM flavor hierarchies. This is made explicit by the approximate $U(2)^5$ symmetry, with leading breaking in the left-handed sector, which we have extensively discussed in the previous sections. In first approximation, this symmetry and symmetry-breaking

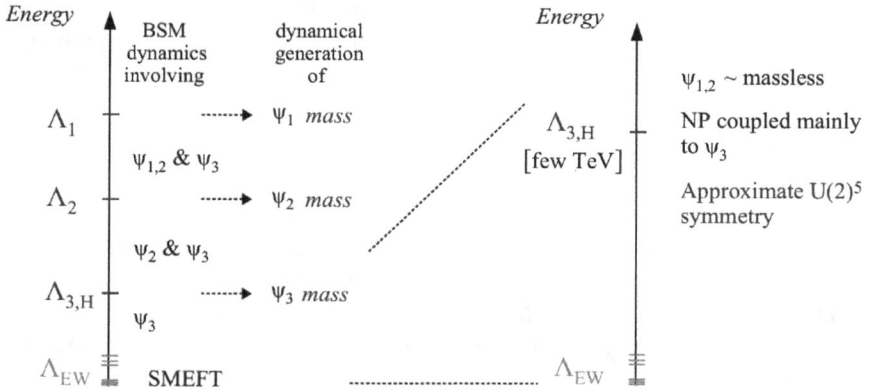

Fig. 4.14: Schematic representation of the possible multi-scale construction at the origin of both flavor anomalies and SM flavor hierarchies.

pattern describes well the anomalies and provides a rational explanation for the structure of the SM Yukawa couplings.

Independently of the nature of the mediators, this flavor pattern can be connected to a multi-scale structure for the underlying dynamics, as schematically indicated in Fig. 4.14. The light families are coupled to NP at higher scales: the small inverse ratios of these scales over the electroweak scale is what determine their effective small Yukawa couplings, and the protection against large deviations from the SM in amplitudes involving only light generations. This pattern, which has been characterised in general terms in Ref. [414, 415], is somehow an old idea, which was proposed before the anomalies appeared, for instance in Ref. [416, 417]. The lowest scale, being around the few TeV and involving the third generation, can be connected to a possible stabilisation of the Higgs sector hence a solution of the electroweak hierarchy problem.

An explicit realization of this multi-scale structure, addressing specifically the $B$-physics anomalies is the three-site PS model proposed in [398]. The basic idea of this construction is that quark-lepton unification a là Pati Salam occurs for all the three families, but for each generation it occurs at a different energy scale. The apparent universality of the SM gauge sector is only a low-energy property. The same is true for the approximate $U(2)^5$ flavor symmetry, which is only due to the high values of the unification scales of the first two families. A schematic view of the symmetry breaking pattern of $PS^3$ down to the SM gauge group is illustrated in Fig. 4.15.

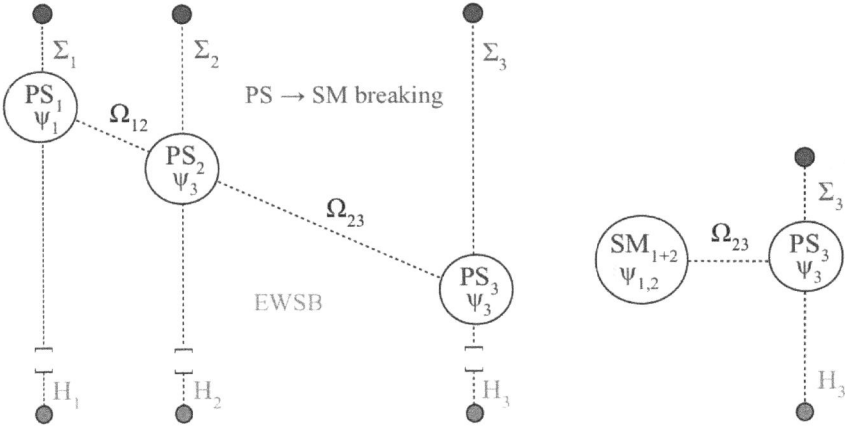

I. High-scale model          II. TeV-scale effective model

Fig. 4.15: Schematic representation of the spontaneous symmetry breaking structure $PS^3 \to SM$ [398, 405]: each dotted line denotes a field (or set of fields) with non-vanishing vacuum expectation values. Long (short) lines qualitatively indicate small (large) vacuum expectation values, respectively.

The three site model of Ref. [398] can in turn can be viewed as a discretized version of a model with an extra compact space-time dimension and PS symmetry in the bulk [405]. In this context, flavor is nothing but a special position in the extra dimension, a topological discontinuity, where a given fermion family is (quasi) localised. The different NP scales identified in the four-dimensional perspective are in one-to-one correspondence to the distances, in the extra dimension, among these different localization points.

These latter ideas are quite speculative and far from being verified. But they are illustrative of the high impact that the *b*-physics anomalies are having in our understanding of the microscopic laws of Nature.

# Chapter 5

# Searches for lepton flavor violation and lepton number violation

## 5.1 Lepton flavor violation

### 5.1.1 *Theoretical considerations*

The evidences of Lepton Flavor Universality violations discussed in chapter 3 naturally open the possibility of observing Lepton Flavor Violation (LFV) in $B$ decays into two differently flavoured charged leptons. In particular, transitions of the type $b \to s(d)\,\tau\mu$ are expected with branching fractions in the range $10^{-9}$–$10^{-4}$, depending from the specific channel, which are within future experimental reach. This fact, which has been pointed out first by Glashow, Guadagnoli, and Lane [343], is confirmed by a large number of explicit models and EFT approaches addressing one or both LFU anomalies (see e.g. [344, 364, 397, 398, 402, 418–425]). Representative predictions of LFV rates for the class of models addressing both anomalies discussed in chapter 4 are shown in Fig. 4.12.

The natural link between LFU and LFV can be understood by general symmetry considerations. As discussed in section 1.4.5, LFU is an accidental symmetry of the SM Lagrangian in the limit of where we neglect the three lepton Yukawa couplings. In such limit, the lepton sector of the SM Lagrangian is invariant under the global symmetry $\mathcal{G}_\ell = U(3)_L \times U(3)_E$, which prevents both LFU violations and LFV. The non-vanishing value of $Y_E$ implies

$$\mathcal{G}_\ell = U(3)_L \times U(3)_E \xrightarrow{Y_E \neq 0} U(1)_e \otimes U(1)_\mu \otimes U(1)_\tau, \qquad (5.1)$$

i.e. a small breaking of LFU and an accidental conservation of lepton flavor. Any other source of breaking of $\mathcal{G}_\ell$, not aligned to $Y_E$ in flavor space, implies LFV. The evidences of LFU violations imply that $\mathcal{G}_\ell$ is broken non only by $Y_E$, but also by some other interaction affecting semileptonic $B$ decays. Beside the rather peculiar limit where this new interaction is aligned to $Y_E$ in flavor space, we should therefore expect also sizable LFV effects in $B$ decays.

An independent breaking of $\mathcal{G}_\ell$, not aligned in flavor space to $Y_E$, is represented by the neutrino mass matrix (see section 1.4.4). Indeed violations of lepton flavor are unambiguously observed in neutrino oscillations. However, neutrino masses are so small that, if they represent the only source of LFV, then the effect is completely negligible in $B$ decays. On the other hand, the recent evidences of LFU violation point to a much larger source of $\mathcal{G}_\ell$ breaking which naturally implies sizable LFV amplitudes in semileptonic $B$ decays.

### 5.1.2 *Experimental searches*

Even prior to the LFU anomalies there were searches for LFV $B$ decays. In particular, in 1993 the CLEO collaboration set the upper limit $\mathcal{B}\left(B^0 \to e^\pm \mu^\mp\right) < 5.9 \times 10^{-6}$ [426] and upper limits on $B^0 \to \mu^\pm \tau^\mp$ and $B^0 \to e^\pm \tau^\mp$, that were superseded in 2004 by the same collaboration [427] obtaining $\mathcal{B}(B^0 \to \mu^\pm \tau^\mp) < 3.8 \times 10^{-5}$ and $\mathcal{B}(B^0 \to e^\pm \tau^\mp) < 1.1 \times 10^{-4}$. These limits were all at 90% confidence level. They were improved upon by the BaBar collaboration [428] that, using hadronic tagging (see section 3.4.2), was able to set $\mathcal{B}(B^0 \to \mu^\pm \tau^\mp) < 2.8 \times 10^{-5}$ and $\mathcal{B}(B^0 \to e^\pm \tau^\mp) < 2.2 \times 10^{-5}$. For a complete list of the early measurements see Ref. [12]. A class of searches where BaBar has made a significant step forward, also using hadronic tagging, is the search for $B^\pm \to h^\pm \tau \ell$ [379]. The corresponding upper limits are listed in Table 5.1.

Table 5.1: 90% CL upper limits obtained by BaBar on LFV decays of the type $B^\pm \to h^\pm \tau \ell$.

| Mode | Upper limit ($\times 10^{-5}$) |
|---|---|
| $B^+ \to K^+ \tau^- \mu^+$ | <4.5 |
| $B^+ \to K^+ \tau^+ \mu^-$ | <2.8 |
| $B^+ \to K^+ \tau^- e^+$ | <4.3 |
| $B^+ \to K^+ \tau^+ e^-$ | <1.5 |
| $B^+ \to \pi^+ \tau^- \mu^+$ | <6.2 |
| $B^+ \to \pi^+ \tau^+ \mu^-$ | <4.5 |
| $B^+ \to \pi^+ \tau^- e^+$ | <7.4 |
| $B^+ \to \pi^+ \tau^+ e^-$ | <2.0 |

### $b \to s(d)\,\tau\mu$ searches by LHCb

In 2020 LHCb decided to search for $B^+ \to K^+\tau^+\mu^-$ in order to see if it were possible to observe this mode using the full available luminosity from the 7, 8 and 13 TeV data. Here they used a novel technique [429] that allows for searches with one missing particle by using the decay $B_{s2}^{*0} \to K^- B^+$ to provide an extra kinematic constraint so they can treat the $\tau^+$ as a missing particle. Actually, they do require the presence of at least one charged track from the $\tau^+$ decay downstream of the $K^+\mu^-$ vertex. The efficiency of this technique is estimated to be about 0.5–1.0% of all $B^+$ decays, thus conquerable to the $B$-factories hadronic tagging efficiency. The result is $\mathcal{B}(B^+ \to K^+\tau^+\mu^-) < 3.9 \times 10^{-5}$ also at 90% CL, somewhat worse that the BaBar limit. This result indicate an interesting potential for LHCb on this type of decay when the $50\,\mathrm{fb}^{-1}$ expected after the Upgrade I will become available.

LHCb also searched for the pure leptonic modes $B_{(s)}^0 \to \tau^\pm\mu^\mp$. Here only $\tau \to \pi\pi\pi\nu$ decays were used, so that the $\tau$ decay point was well defined. Then using kinematically constraints enhanced with a neural network selection, the following 95% CL have been set:

$$\mathcal{B}(B^0 \to \tau^\pm\mu^\mp) < 1.4 \times 10^{-5} \quad \text{and} \quad \mathcal{B}(B_s^0 \to \tau^\pm\mu^\mp) < 4.2 \times 10^{-5}.$$
$$(5.2)$$

This is the best upper limit on the $B^0$ decay and the only one so far available on the $B_s^0$ decay. More data will allow LHCb to enter the interesting region relevant to the LFU anomalies show in Figure 4.12.

### $b \to s(d)\,\mu e$ searches by LHCb

LFV amplitudes involving only light leptons are not expected to be within the experimental reach in models addressing the LFU anomalies, but are of course interesting in more general NP frameworks. LHCb made a search for $B^+ \to K^+\mu^\pm e^\mp$ using $3\,\mathrm{fb}^{-1}$ of Run 1 data. They have been able to set the limits $\mathcal{B}(B^+ \to K^+\mu^-e^+) < 9.5 \times 10^{-9}$ and $\mathcal{B}(B^+ \to K^+\mu^+e^- < 8.8 \times 10^{-9})$, in both cases at 95% CL [430]. These results improves upon the corresponding limits previously obtained by BaBar [431].

LHCb also searched for the helicity suppressed decays $B_{(s)}^0 \to e^\pm\mu^\mp$. The limits obtained using $3\,\mathrm{fb}^{-1}$ of Run 1 data are: $\mathcal{B}(B^0 \to e^\pm\mu^\mp) < 1.3 \times 10^{-9}$ and $\mathcal{B}(B_s^0 \to e^\pm\mu^\mp) < 6.3 \times 10^{-9}$, both at the 95% CL [378].

## 5.2 Lepton number violation

Total lepton number is another accidental symmetry of the SM Lagrangian. Similarly to lepton flavor, a possible violation of total lepton number in $B$ decays would provide a clear evidence of physics beyond the SM. In such processes, total Lepton Number Violation (LNV) could manifest itself via a decay into two charged leptons with the same charge.

The neutrino mass matrix is likely to be described by a dimension-five operator violating total lepton number (see section 1.4.4). However, so far there is no unambiguous evidence of the Majorana-type nature of neutrino masses (see appendix A) and of a violation of total lepton number. In the neutrino sector, the evidence of LNV could be provided by the observation of neutrinoless double-$\beta$ decay ($0\nu\beta\beta$). The double-$\beta$ decay of two nucleons within a single nucleus, with the emission of two electrons and two neutrinos, is a rare second-order weak processes which has been observed in several cases. What has not been observed yet is the same process without the emission of neutrinos, which is possible only if neutrinos have a Majorana mass and total lepton number is violated, as illustrated in Fig. 5.1(a). These decays are currently being investigated. Limits on $0\nu\beta\beta$ lifetimes of $1.8 \times 10^{26}$ years has been set using $^{76}$Ge decays by the GERDA collaboration [432]. Similarly, lifetimes shorter than $3.5 \times 10^{25}$ years have been excluded by the EXO collaboration using $^{136}$Xe [433]. More experiments of this type are in construction.

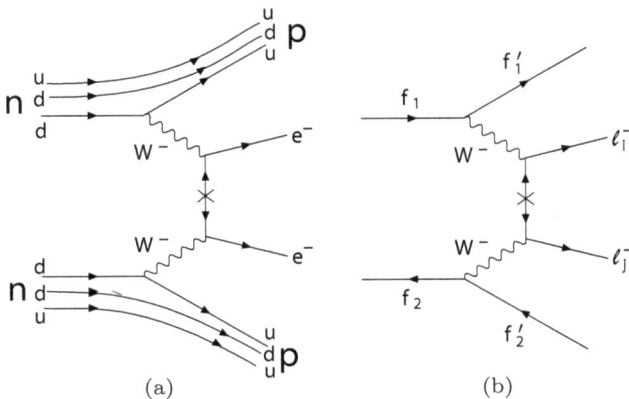

Fig. 5.1: (a) Feynman diagram for the $0\nu\beta\beta$ decay, and (b) analogous quark-level diagram for a LNV charged-meson decay. Here $f_1$ indicates an $s$, $c$, or $b$ quark, $f_2$ indicates a lighter antiquark, and the "×" indicates the Majorana mass term.

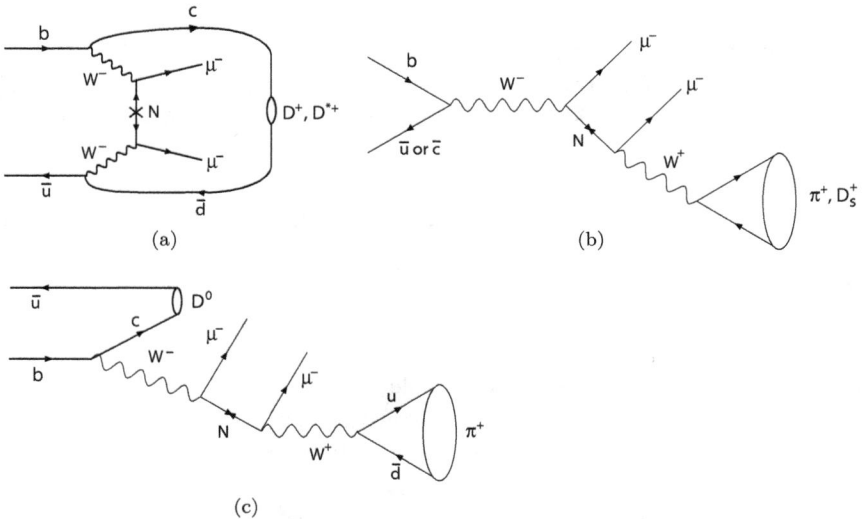

Fig. 5.2: Feynman diagrams for LNV $B$ decays mediated by off-shell and on-shell Majorana mass term for standard or exotic neutrinos ($N$): (a) $B^- \to D^{(*)+}\mu^-\mu^-$, (b) $B^-_{(c)} \to \pi^+(D^+_s)\mu^-\mu^-$, and (c) $B^- \to D^0\pi^+\mu^-\mu^-$.

A conceptually similar LNV transition is possible in meson decays, via the quark-level diagram in Fig. 5.1(b). An example of a specific hadronic process is shown in Fig. 5.2(a). Here the $c$ and $\bar{d}$ form a $D^+$ or a $D^{*+}$ meson. This process is mediated by a virtual Majorana neutrino. The Belle collaboration performed searches for Majorana neutrinos in $B^- \to D^+\ell^-\ell'^-$ decays using their full data sample. The LHCb collaboration searched for both $B^- \to D^+\mu^-\mu^-$ and $B^- \to D^{*+}\mu^-\mu^-$ decays [434], using $0.4\,\mathrm{fb}^{-1}$ of Run 1 data. The limits are shown in Table 5.2. If we interpret these limits into bounds on the Majorana masses for the neutrinos, they turn out to be several orders of magnitude weaker compared to those extracted from $0\nu\beta\beta$. However, these searches have an intrinsic interest in more exotic extensions of the SM, with more sources of LNV, not necessarily connected to the light neutrino masses.

In more exotic extensions of the SM, with exotic heavy neutrinos ($N$) mixing with ordinary neutrinos, other LNV processes can occur, such as those illustrated in Fig. 5.2(b) and 5.2(c). In this case the heavy exotic neutrinos can be produced on-shell and decay as $N \to \pi^+\mu^-$ [436, 437]. For $B^- \to \pi^+\mu^-\mu^-$, we can access the Majorana mass region from approximately 260 MeV to 5150 MeV. This search was first performed

Table 5.2: Upper limits on branching fractions of $B^- \to D^{(*)+}\ell^-\ell^-$ decays.

| Reaction | Exp. | Upper Limit $\times 10^{-6}$ |
|---|---|---|
| $B^- \to D^+ e^- e^-$ | Belle [435] | <2.6 |
| $B^- \to D^+ e^- \mu^-$ | Belle [435] | <1.8 |
| $B^- \to D^+ \mu^- \mu^-$ | Belle [435] | <1.0 |
| $B^- \to D^+ \mu^- \mu^-$ | LHCb [434] | <0.69 |
| $B^- \to D^{*+} \mu^- \mu^-$ | LHCb [434] | <3.6 |

by Mark-II [438] and then by CLEO [439]. With 2010 data, LHCb also performed similar analysis [440] reporting an upper limit of $4.4 \times 10^{-8}$.

The above upper limits are obtained assuming that the heavy neutrino has a short lifetime and decays at the $B$ decay point. However, if the $N$ is longer lived, these limits do not necessarily apply. Belle, using their full data sample [441], and LHCb [442], using $3\,\text{fb}^{-1}$ of Run 1 data, analyzed the $B^- \to \pi^+ \mu^- \mu^-$ decay for different $N$ lifetimes. No signals have been observed for lifetimes up to 200 ps, above which the efficiency becomes very small. The branching ratio upper limits from LHCb as a function of $N$ mass and lifetime are shown in Fig. 5.3.

Model dependent upper limits on the coupling of a single fourth-generation Majorana neutrino to muons, $|V_{\mu 4}|$, for each value of the mass of $N$ ($m_N$) can be extracted using the following formula from Atre *et al.* [436]

$$\mathcal{B}(B^- \to \pi^+ \mu^- \mu^-) = \frac{G_F^4 f_B^2 f_\pi^2 m_B^5}{128\pi^2 \hbar} |V_{ub}V_{ud}|^2 \tau_B \left(1 - \frac{m_N^2}{m_B^2}\right) \frac{m_N}{\Gamma_N} |V_{\mu 4}|^4. \tag{5.3}$$

A key ingredient to extract the bound is the value of $\Gamma_N$, which in turn depend on $m_N$ and $|V_{\mu 4}|$. Using the QCD-based estimates of $\Gamma_N$ [443], which should replace the *ad hoc* estimates used in the original experimental publications, leads to the bounds on $|V_{\mu 4}|$ shown in Fig. 5.4.

Finally, it is worth mention the process $B^- \to D^0 \pi^+ \mu^- \mu^-$, suggested in Ref. [444], corresponding to Fig. 5.2(c), which could benefit from a higher branching ratio compared to $B^- \to \pi^+ \mu^- \mu^-$ due to a non-suppressed hadronic matrix element. Here the accessible mass region for Majorana neutrino mass is between 260 MeV and 3300 MeV. Using $0.4\,\text{fb}^{-1}$ of Run 1 data, LHCb [434] set the upper limit $\mathcal{B}(B^- \to D^0 \pi^+ \mu^- \mu^-) < 10^{-6}$ at 95% CL.

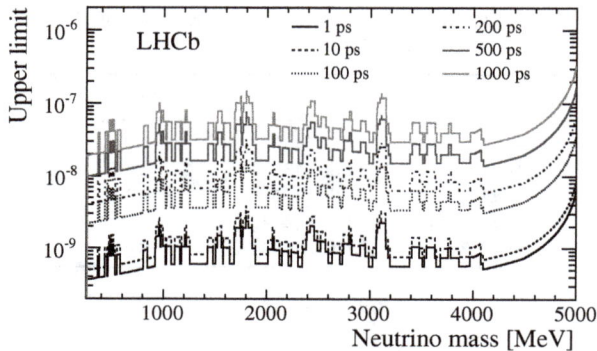

Fig. 5.3: Upper limits on $\mathcal{B}(B^- \to \pi^+\mu^-\mu^-)$ at 95% C.L. as a function of $m_N$, in 5 MeV intervals, for specific values of $\tau_N$ from [442].

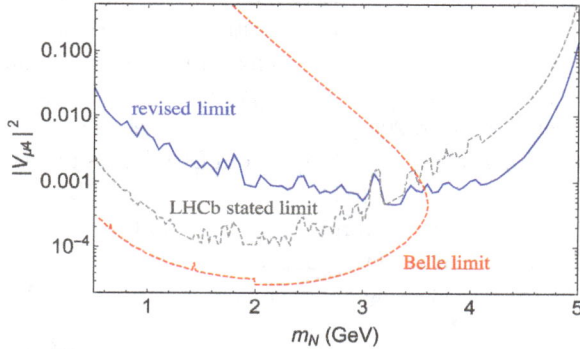

Fig. 5.4: Upper limits on $|V_{\mu 4}|^2$ at 95% confidence level from the LHCb and Belle experiments as interpreted in Ref. [443]. The (grey) dashed line shows the limit from the LHCb paper. The (blue) solid line shows the limit using the corrected model. For comparison, the lower (red) dashed line shows the revised limit from Belle. In all cases, it is assumed that $V_{e4} = V_{\tau 4} = 0$.

In many realistic extensions of the SM these limits are still far from the most interesting parameter-space regions [445, 446]. However, as stated above, these searches have an intrinsic interest providing a unique probe of a so far unchallenged accidental symmetry of the SM. It is also worth noting that all these results can be significantly updated with the large amount of integrated luminosity that LHCb expects to collect in the future.

# Chapter 6

# Conclusions

## 6.1 Summary about the evidence of New Physics and theoretical insights

Decays of hadrons containing $b$ quarks have been an essential ingredient in our understanding of the mechanism of flavor mixing within Standard Model. Even more importantly, they are also a formidable tool to search for physics beyond the Standard Model. In the last few years a series of anomalies, i.e. small but non-negligible deviations from the SM predictions, have been accumulating in the experimental study of $b$-hadron decays. They are naturally grouped into two categories. The first one includes all processes based on the quark-level transition $b \to s\ell^+\ell^-$, which occur at the loop level within the SM (see Fig. 3.1). The other consists of transitions of the type $b \to c\tau\bar{\nu}$, which are governed by a tree level diagram within the SM (see Fig. 1.1(a)). There is no obvious link among the two, but they both indicate a possible violation of Lepton Flavor Universality: a $\mu/e$ non-universality in the $b \to s\ell^+\ell^-$ case, and a special role for the $\tau$ in the $b \to c\tau\bar{\nu}$ case. These are effects which cannot be attributed to theoretical uncertainties within the SM, and which could indicate a common origin of the two anomalies in extensions of the SM.

In $b \to s\ell^+\ell^-$ transitions the NP signals are numerous and coherent. The most striking deviations from the SM predictions are in the LFU ratios $R_{K^{(*)}}$, defined in Eq. (3.1), which exhibit three independent deviations in different dilepton regions and different final states, ranging from $2.5\sigma$ (low $q^2$ bin in $R_{K^*}$) up tp $3.1\sigma$ (single $q^2$ bin in $R_K$). These are accompanied by a series of other anomalies in dilepton spectra and angular distributions. When combined, current $b \to s\ell^+\ell^-$ data provide a strong indication of physics beyond the SM. Using a very conservative approach, the significance of non-standard interactions is around the $4\sigma$ level. The significance raises to $4.7\sigma$ restricting the attention only to motivated NP frameworks, and estimating theory errors in a very conservative manner. The data neatly point to the motivated framework of leptoquark-mediated semileptonic operators respecting an approximate $U(2)^5$ flavor symmetry, but other options, such as a non-universally coupled $Z'$ boson, cannot be excluded at this stage.

In the $b \to c\tau\bar{\nu}$ the case the two LFU ratios $R_D$ and $R_{D^*}$, defined in Eq. (3.18), are the only anomalies. Their status is summarized by the HFLAV fit shown in Fig. 3.29 [74]. There the combined deviation from the SM is estimated to be at the $3.1\sigma$ level of significance, after averaging over all the results. The experimental situation, however, remains unclear. While all results are above the SM expectations, the error bars on most are large enough so that each of the results are consistent within $2\sigma$ with the SM, if taken separately. The only exception is the 2012 result from BaBar. These are very difficult measurements, both at $e^+e^-$ $B$-factories and at LHCb. As we have discussed in detail in chapter 3, the most serious concern is the estimate of the backgrounds (in particular those induced by excited $D$ states, misidentified as $D$ or $D^*$, or semiletonic $D$ decays from double charm production). While the experiments may have covered these problems with sufficient systematic uncertainties, if all results are above the SM predictions, we cannot exclude that this effect is accounted for by a common systematic uncertainty due to correlated backgrounds. Having said this, these are clearly very interesting results, especially when taking into account the independent indication of LFU violation coming from the cleaner $b \to s\ell^+\ell^-$ transitions.

Overall, is too early to draw definite conclusions. It is possible that only one set of anomalies turns out to be a true deviations from the SM and the other not, the $b \to s\ell^+\ell^-$ system being clearly favored given the higher significance and overall consistency. If this is the case, many theoretical interpretations are possible. However, if both anomalies are confirmed to be true BSM manifestations, a very interesting and quite distinct pattern seem to emerge, as discussed at length in chapter 4. First, the underlying dynamics should have a hierarchical flavor structure similar to the one observed in the SM Yukawa couplings, pointing to a possible common origin of flavor anomalies and flavor hierarchies. Second, the combination of all data (not only the anomalies, but also the numerous stringent tests of the SM at low and high energies) point quite clearly to one class of mediators: the $U_1$ vector leptoquark (see table 4.4). On the more speculative side, this setup seems to point toward a unification of quark and leptons in a rather unconventional way, i.e. family by family at different energy scales. This would represent a major shift in our understanding of fundamental interactions.

Beside this fascinating but still premature conclusion, what clearly emerges from present observations is the need of more data to shed more light on the origin of the anomalies so far observed. In most cases, theoretical models addressing the anomalies point to a series deviations from the SM in other $B$ decays not far from present bounds. Interestingly enough, a non-negligible amount of data relevant to these studies already collected at LHCb still need to be analysed. Most importantly, a significant increase in the $b$-physics data set at both $e^+e^-$ $B$-factories and hadron colliders are expected in the near future, as we discuss below.

## 6.2   Future prospects

Data so far analysed in the $b \to s\ell^+\ell^-$ system represent only a fraction of those collected by LHCb during Run 1 and Run 2. It is not obvious that the $5\sigma$ discovery level will be reached with the data already recorded, but this is not impossible. We recall that the present results on $R_{K^*}$ are based on Run 1 statistics only. Moreover, LFU ratios in other channels, such as $R_\phi$ in $B_s^0 \to \phi\ell^+\ell^-$ decays, and $R_\Lambda$ in $\Lambda_b^0 \to \Lambda\ell^+\ell^-$, are in progress. These have substantially less statistics than $R_{K^{(*)}}$, but can contribute to the overall significance of the anomalies if the results will be coherent with present findings. Last but not least, other interesting measurements can be expected from the leading channels, in particular the LFU analysis on the angular distribution in $B \to K^*\ell^+\ell^-$ is particularly promising. Given the present significance of $b \to s\ell^+\ell^-$ anomalies is already quite high, the addition of this extra information promises to be quite interesting.

Beside data already collected, a significant increase in LHCb statistics is expected when the installation of the Upgrade I will be completed and the LHC will start again to produce collisions in 2022. The expected accumulated luminosity as a function of time is shown in Fig. 6.1. A further upgrade is being planned after the long shut down 4 (LS4) of the LHC accelerator complex. This is not approved yet, but LHCb is in the processes of responding to the request made by the LHCC (the CERN review committee for the LHC program) of a framework technical design report for Upgrade II.

So far (i.e. during runs 1 and 2), LHCb took data at an instantaneous luminosity of $4 \times 10^{32} \, \mathrm{cm}^{-1}\mathrm{s}^{-1}$ using a fast hardware trigger that was

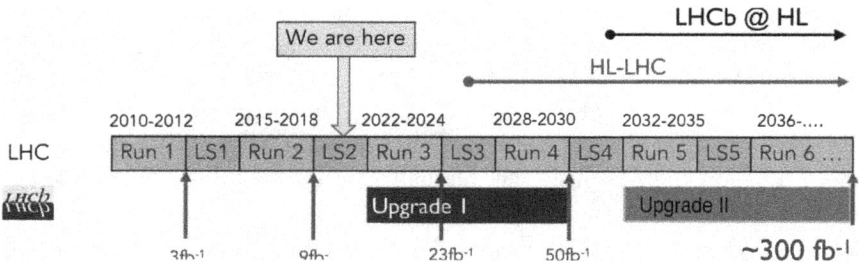

Fig. 6.1: Long term schedule for LHCb Upgrades I and II and the approximate luminosity accumulation as a function of time. Uncertainties due to the Sars-covid-2 pandemic are still present.

about 50% efficient on fully hadronic $b$ decays, with the exact number dependent on decay mode, and $\sim$90% efficient on dimuon final states. LHCb runs in very stable conditions with the instantaneous luminosity being held constant throughout the run. This is achieved by initially displacing the two proton beams so they are not colliding head on and moving them closer together when the number of protons decreases. This is very advantageous as the backgrounds from multiple $pp$ collisions in a bunch stays constant thus calibrations can be averaged over the data sample. In the Upgrade I phase the hardware trigger is eliminated and the instantaneous luminosity is increased to $2 \times 10^{33}\,\mathrm{cm}^{-1}\,\mathrm{s}^{-1}$, providing a factor of 5–10 increase in events per unit time depending on the final states. In Upgrade II the instantaneous luminosity is further increased to $2 \times 10^{34}\,\mathrm{cm}^{-1}\,\mathrm{s}^{-1}$ providing, in principle, another order of magnitude increase in the data rate. Unfortunately, luminosity levelling cannot be fully implemented in Upgrade II due to the higher collision rate.

The Belle II luminosity projection is shown in Fig. 6.2 [447]. Belle II should accumulate about as much data by 2022 as Belle gathered over its lifetime. Then there will a long shutdown to install new equipment, including part of the vertex detector. Luminosity gains are expected until 2026 when there is a major upgrade of the interaction region to allow for a maximum value of instantaneous luminosity of $6.5 \times 10^{35}\mathrm{cm}^{-1}\mathrm{s}^{-1}$. When comparing the LHCb and Belle II luminosities, it is important to recall that the $B$ meson cross-section at the $\Upsilon(4S)$ is $2 \times 1.1\,\mathrm{nb}$, while the $b$-hadron cross-section in the LHCb acceptance at $13\,\mathrm{TeV}$ is $144\,\mu\mathrm{b}$ [30].

In both programs (LHCb and Belle II), the projected increases in accumulated data are factors of 30–50. This would allow for major advancements in the field, which is particularly exciting given the present discrepancies

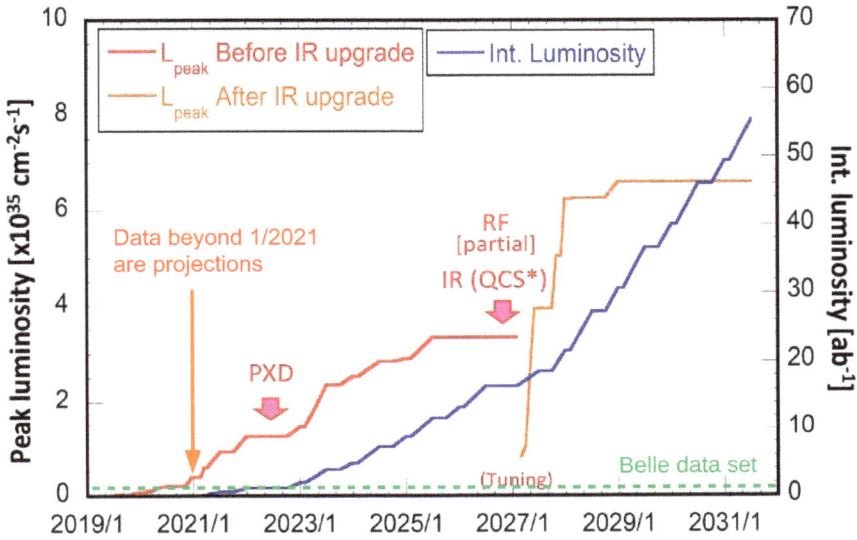

Fig. 6.2: Projected luminosity accumulation for Belle II (blue, right side scale) and instantaneous luminosity (red, left side scale) as a function of time. The dashed (green) line shows the total Belle data sample.

between data and SM predictions. If the present LFU anomalies are a true signals of physics beyond the SM, this large increase in statistics will be essential to further explore this phenomenon. As we pointed out in chapter 4, these anomalies cannot be an isolated: they are likely to be accompanied by a series of other interesting non-standard effects at low energies and, especially, in $b$-quark decays. These include LFV decays of the type $b \to s(d)\tau\mu$, enhanced rates for processes of the type $b \to s\tau^+\tau^-$ and $b \to s\nu\bar{\nu}$, and others. The future programs of LHCb and Belle II should be sensitive to significant fractions of anomaly based predictions of these and other branching fractions.

Of course one must take into account the famous adage: "It is difficult to make predictions, especially about the future." At this stage we cannot exclude yet that none of the anomalies will be confirmed. Still, the future $b$-physics programs remain a very interesting opportunity to probe in depth the validity of the SM. It is indeed also possible that new physics will reveal itself in an entirely different and as up to now unheralded reaction, or even in the precision measurements discussed in chapter 2.

# Appendix A
# The SM Lagrangian

The three pieces of the SM Lagrangian, decomposed as in Eq. (1.1), are

$$\mathcal{L}_{\text{gauge}} = -\frac{1}{4g_s} \sum_{a=1}^{8} G^{a\mu\nu} G^a_{\mu\nu} - \frac{1}{4g_2} \sum_{a=1}^{3} W^{a\mu\nu} W^a_{\mu\nu} - \frac{1}{4g_1} B^{\mu\nu} B_{\mu\nu}$$

$$+ \sum_{a=1}^{3} \sum_{\psi=Q,L,u,d,e} \bar{\psi}^a i \gamma_\mu D^\mu \psi^a \tag{A.1}$$

$$\mathcal{L}_{\text{Higgs}} = D^\mu H^\dagger D_\mu H - \frac{\lambda}{4} (H^\dagger H)^2 + \mu^2 (H^\dagger H) \tag{A.2}$$

$$-\mathcal{L}_{\text{Yukawa}} = (Y_D)_{ab} \, \overline{Q}^a_L H d^b_R + (Y_U)_{ab} \, \overline{Q}^a_L H_c u^b_R + (Y_E)_{ab} \, \overline{L}^a_L H e^b_R + \text{h.c.,} \tag{A.3}$$

where the summation over Lorentz indices is left implicit. Here $g_i$ denote the three gauge couplings ($g_s \equiv g_3$ and $g_1 \equiv g_Y$) and the covariant derivative assumes the form

$$D_\mu = \partial_\mu - ig_3 \sum_{a=1}^{8} \hat{T}^a_{[3]} G^a_\mu - ig_2 \sum_{a=1}^{3} \hat{T}^a_{[2]} W^a_\mu - ig_1 \hat{Y} B_\mu, \tag{A.4}$$

where $\hat{T}^a_{[3]}$, $\hat{T}^a_{[2]}$, and $\hat{Y}$ denote the generators of $SU(3)$, $SU(2)$, and hypercharge respectively, in a given representation.

The parameters of the Higgs potential, $\mu > 0$ and $\lambda > 0$, are such that the Higgs field develops a non-vanishing vacuum expectation value. Without loss of generality, the latter can be chosen as

$$\langle H \rangle = \frac{1}{\sqrt{2}} \begin{pmatrix} 0 \\ v \end{pmatrix}. \tag{A.5}$$

The non vanishing value of $\langle H \rangle$ leads to the spontaneous symmetry breaking of the electroweak symmetry $SU(2)_L \times U(1)_Y \rightarrow U(1)_Q$. The QED coupling $e$, associated to the unbroken $U(1)_Q$ group, can be expressed as

$$e = \frac{g_1 g_2}{g_1^2 + g_2^2} \equiv g_2 \sin \theta_W \equiv g_1 \cos \theta_W. \tag{A.6}$$

The spontaneous symmetry breaking of the electroweak symmetry leads to non-vanishing masses for the $W^{\pm}$ and $Z$ fields. At the tree level, their masses can be expressed as

$$m_W^2 = \frac{g_2^2}{4} v^2, \quad m_Z^2 = \frac{g_1^2 + g_2^2}{4} v^2, \tag{A.7}$$

in terms of the gauge couplings and the Higgs vacuum expectation value. The precise value of $v$ is determined by the Fermi coupling, defined in Eq. (1.22), via the relation

$$v = \left( \sqrt{2} G_F \right)^{-1/2} \approx 246 \, \text{GeV}. \tag{A.8}$$

The conjugate Higgs field appearing in $\mathcal{L}_{\text{Yukawa}}$ is defined as $H_c = i\sigma_2 H^{\dagger}$.

For more details, and in particular explicit expressions of the interaction terms in the mass-eigenstate basis of the different SM fields, see e.g. Ref. [448–452].

## A.1 Dirac versus Majorana masses

Within the SM, fermions have a Dirac-type mass (quarks and charged leptons) or are massless (left-handed neutrinos). Once the Higgs field acquires its vacuum expectation value, an effective Lorentz-invariant bilinear coupling between a pair of fermions with different helicity (a left-handed and a right-handed field) is generated:

$$\text{Dirac mass}: \quad \bar{\psi}_L \psi_R \quad [\text{invariant under } \psi_{L(R)} \rightarrow e^{i\alpha} \psi_{L(R)}]. \tag{A.9}$$

The gauge symmetry, the fermion content, and the request of considering only renormalizable operators (hence operators of dimension $d \leq 4$) prevent to write any Majorana-type mass, i.e. a Lorentz-invariant bilinear coupling between two identical fermions [453]:

$$\text{Majorana mass}: \quad \bar{\psi}_L^c \psi_L \quad \text{or} \quad \bar{\psi}_R^c \psi_R \quad [\text{not invariant under } \psi_{L(R)} \rightarrow e^{i\alpha} \psi_{L(R)}], \tag{A.10}$$

where $\psi^c = i\gamma_2\psi^*$. A Majorana mass is possible, for the left-handed neutrinos, if considering $d = 5$ operators, as shown in Eq. (1.33). If the model is extended adding one or more right-handed neutrinos, which would be neutral with respect to the whole SM gauge group, these field could have a Majorana mass term already at $d = 3$. For more details about possible neutrino mass terms see e.g. Ref. [454].

# Appendix B

# Flavor symmetries

## B.1  $U(3)^5$ and minimal flavor violation

The largest group of flavour-symmetry transformations compatible with the gauge symmetries of the SM Lagrangian is [455, 456]

$$U(3)^5 = U(3)_L \otimes U(3)_Q \otimes U(3)_e \otimes U(3)_u \otimes U(3)_d = SU(3)^5 \otimes U(1)^5, \quad \text{(B.1)}$$

where $\{L, Q, e, u, d\}$ denote the five independent types of SM fermions with different gauge quantum numbers reported in Table 1.1. Within the SM, the Yukawa couplings $(Y_{E,U,D})$ are the only source of breaking of $U(3)^5$. They break this global symmetry as follows

$$\begin{cases} SU(3)^5 \xrightarrow{Y_{E,U,D} \neq 0} U(1)_{e-\mu} \otimes U(1)_{\tau-\mu} \\ U(1)^5 \qquad\qquad U(1)_B \otimes U(1)_L \otimes U(1)_Y \end{cases} \quad \text{(B.2)}$$

where we separated explicitly flavour-universal and flavour-non-universal subgroups. The three unbroken flavour-universal $U(1)$ groups are baryon number, lepton number, and hypercharge.[1]

The Minimal Flavor Violation (MFV) hypothesis is the assumption that the SM Yukawa couplings are the only sources of $U(3)^5$ breaking not only in the SM, but also beyond the SM [157]. To construct operators compatible with this hypothesis, $Y_{E,U,D}$ are interpreted as $U(3)^5$ spurion fields with the following transformation properties:

$$Y_E = (\mathbf{3}, \mathbf{1}, \bar{\mathbf{3}}, \mathbf{1}, \mathbf{1}), \quad Y_U = (\mathbf{1}, \mathbf{3}, \mathbf{1}, \bar{\mathbf{3}}, \mathbf{1}), \quad Y_D = (\mathbf{1}, \mathbf{3}, \mathbf{1}, \mathbf{1}, \bar{\mathbf{3}}). \quad \text{(B.3)}$$

---

[1] The two flavour-non-universal $U(1)$ subgroups left unbroken by the Yukawa couplings are arbitrary combinations of the two diagonal generators of $SU(3)_{e+L}$, namely the vectorial subgroup of $SU(3)_e \otimes SU(3)_L$.

Within the effective-theory approach to physics beyond the SM discussed in section 1.4.4 we can state that the SMEFT satisfies the criterion of MFV if all higher dimensional operators, constructed from SM and the $Y_{E,U,D}$ (treated as spurions), are invariant under $U(3)^5$. Then, for instance, the $d = 6$ term $(\bar{Q}_L^1 \gamma^\mu Q_L^2)(\bar{L}_L^a \gamma^\mu L_L^a)$ is not compatible with the MFV criterion, while the $d = 6$ term $[\bar{Q}_L^a (Y_U^\dagger Y_U)_{ab} \gamma^\mu Q_L^b](\bar{L}_L^c \gamma^\mu L_L^c)$ is allowed.

In principle, the spurions can appear with arbitrary powers both in the renormalizable ($d = 4$) part of the Lagrangian and in the dimension-six effective operators. However, via a suitable redefinition of both fermion fields and spurions, we can always put the $d = 4$ Lagrangian to its standard expression in Eq. (1.2), namely we can always identify the spurions with the SM Yukawa couplings. This implies we can always choose a flavour basis where the spurions are completely determined in terms of fermion masses and the CKM matrix. A particular useful basis to evaluate processes with external down-type quarks, such as $b \to s$ transitions or $B_{s,d}$ mixing, is the down-quark mass-eigenstate basis, where

$$Y_E = \text{diag}(y_e, y_\mu, y_\tau), \quad Y_U = V_{\text{CKM}}^\dagger \times \text{diag}(y_u, y_c, y_t),$$
$$Y_D = \text{diag}(y_d, y_s, y_b). \tag{B.4}$$

As far as the higher-dimensional operators are concerned, a key observation is that all the eigenvalues of the Yukawa matrices are small, but for $y_t$, and that the off-diagonal elements of the CKM matrix are very suppressed. Working in the basis defined by Eq. (B.4), we can thus write

$$\left[Y_U (Y_U)^\dagger\right]_{i \neq j}^n \approx y_t^{2n} V_{ti}^* V_{tj}. \tag{B.5}$$

As a consequence, including high powers of the the Yukawa matrices in the higher-dimensional operators of the SMEFT (see section 1.4.4) amounts only to a redefinition of the overall factor in (B.5). This implies in particular that the leading $\Delta F = 2$ and $\Delta F = 1$ FCNC amplitudes get exactly the same CKM suppression as in the SM:

$$\mathcal{A}(d^i \to d^j)_{\text{MFV}} = (V_{ti}^* V_{tj}) \mathcal{A}_{\text{SM}}^{(\Delta F=1)} \left[1 + a_1 \frac{16\pi^2 M_W^2}{\Lambda^2}\right],$$
$$\mathcal{A}(M_{ij} - \bar{M}_{ij})_{\text{MFV}} = (V_{ti}^* V_{tj})^2 \mathcal{A}_{\text{SM}}^{(\Delta F=2)} \left[1 + a_2 \frac{16\pi^2 M_W^2}{\Lambda^2}\right], \tag{B.6}$$

where $\mathcal{A}_{\text{SM}}^{(i)}$ denote the SM loop amplitudes and $a_i$ are $\mathcal{O}(1)$ parameters. The $a_i$ depend on the specific operator considered but are flavor independent.

## B.2 The $U(2)^5$ symmetry and its minimal breaking

The $U(2)^5$ flavor symmetry is the subgroup of the $U(3)^5$ global symmetry that, by construction, distinguishes the first two generations of fermions from the third one [297–299]. For each set of SM fermions with the same gauge quantum numbers, the first two generations form a doublet of a given $U(2)$ subgroup, whereas the third one transforms as a singlet. The five independent flavor doublets are denoted $Q, L, U, D, E$ and the flavor symmetry decomposes as

$$U(2)^5 = U(2)_Q \otimes U(2)_L \otimes U(2)_U \otimes U(2)_D \otimes U(2)_E. \qquad \text{(B.7)}$$

In the limit of unbroken $U(2)^5$, only third-generation fermions can have non-vanishing Yukawa couplings, which is an excellent first-order approximation for the SM Lagrangian.

A set of symmetry-breaking terms sufficient to reproduce the complete structure of the SM Yukawa couplings is [297]

$$V_q \sim (\mathbf{2}, \mathbf{1}, \mathbf{1}, \mathbf{1}, \mathbf{1}), \quad V_\ell \sim (\mathbf{1}, \mathbf{2}, \mathbf{1}, \mathbf{1}, \mathbf{1}),$$
$$\Delta_{u(d)} \sim (\mathbf{2}, \mathbf{1}, \bar{\mathbf{2}}(\mathbf{1}), \mathbf{1}(\bar{\mathbf{2}}), \mathbf{1}), \quad \Delta_e \sim (\mathbf{1}, \mathbf{2}, \mathbf{1}, \mathbf{1}, \bar{\mathbf{2}}). \qquad \text{(B.8)}$$

By construction, $V_{q,\ell}$ are complex two-vectors and $\Delta_{e,u,d}$ are complex $2 \times 2$ matrices. In terms of these spurions, we can express the Yukawa couplings as

$$Y_E = y_\tau \begin{pmatrix} \Delta_e & x_\tau V_\ell \\ 0 & 1 \end{pmatrix}, \quad Y_U = y_t \begin{pmatrix} \Delta_u & x_t V_q \\ 0 & 1 \end{pmatrix}, \quad Y_D = y_b \begin{pmatrix} \Delta_d & x_b V_q \\ 0 & 1 \end{pmatrix},$$
$$\text{(B.9)}$$

where $y_{\tau,t,b}$ and $x_{\tau,t,b}$ are free complex parameters expected to be of order $\mathcal{O}(1)$.

The set in Eq. (B.9) is minimal in terms of independent $U(2)^5$ structures (at least as far as the quark sector is concerned), and leads to spurions which are small and hiearchical in size. Contrary to the MFV framework, in this case we cannot determine completely the spurions in terms of SM parameters. However, we can constrain their size requiring no tuning in the $\mathcal{O}(1)$ parameters. In particular, from the $2 \leftrightarrow 3$ mixing in the CKM matrix we deduce $|V_q| = O(|V_{cb}|)$, while light quark and lepton masses imply $|\Delta_{u,d,e}|_{ij} \ll |V_q|$. We have no unambiguous constraints about the size of $V_\ell$. Assuming a common structure for the three Yukawa couplings, as suggested by the similar hierarchies observed in the eigenvalues, it is natural to assume $|V_\ell| \sim |V_q|$. The assumption that $V_{q,\ell}$ are the leading

$U(2)^5$–breaking spurions ensures a suppression of flavor-violating terms in the quark sector, via higher-dimensional operators, as effective as the one implied by the MFV hypothesis.

**Yukawa couplings and spurion strucutre**

We define as interaction basis the flavor basis in $U(2)^5$ space where $V_{q,\ell} = |V_{q,\ell}| \times \vec{n}$, with $\vec{n} = (0,1)^{\mathsf{T}}$, and $\Delta^\dagger_{u,d,e}\Delta_{u,d,e}$ are diagonal. After the $U(2)^5$ is broken as in Eq. (B.9), the residual flavor symmetry implies that the Yukawa matrices in the interaction basis can be written in the following form [323]

$$Y_U = |y_t| \begin{pmatrix} U_q^\dagger O_u^{\mathsf{T}} \hat{\Delta}_u & |V_q| |x_t| e^{i\phi_q} \vec{n} \\ 0 & 1 \end{pmatrix}, \quad Y_D = |y_b| \begin{pmatrix} U_q^\dagger \hat{\Delta}_d & |V_q| |x_b| e^{i\phi_q} \vec{n} \\ 0 & 1 \end{pmatrix},$$

$$Y_E = |y_\tau| \begin{pmatrix} O_e^{\mathsf{T}} \hat{\Delta}_e & |V_\ell| |x_\tau| \vec{n} \\ 0 & 1 \end{pmatrix}, \tag{B.10}$$

where $\hat{\Delta}_{u,d,e}$ are $2 \times 2$ diagonal positive matrices, $O_{u,e}$ are $2 \times 2$ orthogonal matrices, and $U_q$ is of the form

$$U_q = \begin{pmatrix} c_d & s_d\, e^{i\alpha_d} \\ -s_d\, e^{-i\alpha_d} & c_d \end{pmatrix}, \quad s_d \equiv \sin\theta_d, \quad c_d \equiv \cos\theta_d. \tag{B.11}$$

The unitary matrices that diagonalize the Yukawa matrices above, defined as $L_f^\dagger Y_f R_f = \mathrm{diag}(Y_f)$ $(f = u, d, e)$, assume the form [323]

$$L_d \approx \begin{pmatrix} c_d & -s_d\, e^{i\alpha_d} & 0 \\ s_d\, e^{-i\alpha_d} & c_d & s_b \\ -s_d\, s_b\, e^{-i(\alpha_d+\phi_q)} & -c_d\, s_b\, e^{-i\phi_q} & e^{-i\phi_q} \end{pmatrix},$$

$$R_d \approx \begin{pmatrix} 1 & 0 & 0 \\ 0 & 1 & \frac{m_s}{m_b} s_b \\ 0 & -\frac{m_s}{m_b} s_b\, e^{-i\phi_q} & e^{-i\phi_q} \end{pmatrix},$$

$$\tag{B.12}$$

$$L_e \approx \begin{pmatrix} c_e & -s_e & 0 \\ s_e & c_e & s_\tau \\ -s_e s_\tau & -c_e s_\tau & 1 \end{pmatrix}, \quad R_e \approx \begin{pmatrix} 1 & 0 & 0 \\ 0 & 1 & \frac{m_\mu}{m_\tau} s_\tau \\ 0 & -\frac{m_\mu}{m_\tau} s_\tau & 1 \end{pmatrix},$$

$$R_u \approx \begin{pmatrix} 1 & 0 & 0 \\ 0 & 1 & \frac{m_c}{m_t} s_t \\ 0 & -\frac{m_c}{m_t} s_t\, e^{-i\phi_q} & e^{-i\phi_q} \end{pmatrix},$$

with $L_u = L_d V_{\text{CKM}}^\dagger$ and $s_t = s_b - V_{cb}$. The parameters $s_d$ and $\alpha_d$ are not free and can be expressed in terms of CKM parameters as follows

$$s_d/c_d = |V_{td}/V_{ts}|, \quad \alpha_d = \arg(V_{td}^*/V_{ts}^*). \qquad \text{(B.13)}$$

The flavor-mixing parameters, or spurion structures, which cannot be determined in terms of SM parameters are:

- *quark sector*: $2 \leftrightarrow 3$ mixing angle $s_b/c_b = |x_b| |V_q|$ and CP-violating phase $\phi_q$;
- *lepton sector*: $2 \leftrightarrow 3$ mixing angle $s_\tau/c_\tau = |x_\tau| |V_\ell|$ and $1 \leftrightarrow 2$ mixing angle $s_e$ (which appears in $O_e$).

## Higher-dimensional operators

The higher-dimensional operators built in terms of the leading $U(2)^5$-breaking spurions $(V_{q,\ell})$ lead to flavor-violating transitions which:

- involve only left-handed fields;
- connect only the $2 \leftrightarrow 3$ sectors in the interaction basis.

The first point above is the reason why in $b \to s\ell^+\ell^-$ transitions we can restrict the attention to left-handed operators in the limit of a minimally-broken $U(2)^5$ flavor symmetry. The fact that $U_d$, and the $2 \times 2$ upper block of $L_d$, are entirely determined in terms of CKM elements, implies that the $2 \leftrightarrow 3$ transitions in the interaction basis correspond to a well-defined direction in the mass-eigenstate basis of the light quarks. In particular, this is the origin of the relation in Eq. (4.21), involving up-type light quarks, and the one in Eq. (4.39), involving down-type light quarks.

The approximate down-alignment in the $2 \leftrightarrow 3$ sector, necessary to satisfy the tight bounds from $B_s$ mixing when addressing the $B$ anomalies, is achieved for $|x_b| \ll 1$. In the exact $2 \leftrightarrow 3$ down-alignment limit, i.e. for $|x_b| \to 0$, the phase $\phi_q$ becomes unphysical and $s_b \to 0$. Hence in this limit there are no unconstrained flavor-mixing parameters in the quark sector.

It is finally worth to note that the hypothesis that the $U(2)^5$ flavor symmetry is broken by the minimal set of spurions in Eq. (B.9) naturally implies lepton flavor violation in charged leptons. This is controlled by the size of $V_\ell$ and $s_e$, which are left unconstrained by the SM Yukawa couplings.

# Appendix C

# Effective $b \to s\ell^+\ell^-$ operators

Employing the decomposition of the $b \to s\ell^+\ell^-$ effective Lagrnagian in Eq. (4.1), the subleading four-quark operators not listed in section 4.2.1 can be defined as

$$Q_3^{(4q)} = (\bar{s}_L \gamma_\mu b_L) \sum_q (\bar{q} \gamma^\mu q), \quad Q_4^{(4q)} = (\bar{s}_L \gamma_\mu T^a b_L) \sum_q (\bar{q} \gamma^\mu T^a q),$$

$$Q_5^{(4q)} = (\bar{s}_L \gamma_\mu \gamma_\nu \gamma_\rho b_L) \sum_q (\bar{q} \gamma^\mu \gamma^\nu \gamma^\rho q), \tag{C.1}$$

$$Q_6^{(4q)} = (\bar{s}_L \gamma_\mu \gamma_\nu \gamma_\rho T^a b_L) \sum_q (\bar{q} \gamma^\mu \gamma^\nu \gamma^\rho T^a q),$$

and

$$Q_7^{(4q)} = (\bar{s}_L \gamma_\mu b_L) \sum_q e_q (\bar{q} \gamma^\mu q), \quad Q_8^{(4q)} = (\bar{s}_L \gamma_\mu T^a b_L) \sum_q e_q (\bar{q} \gamma^\mu T^a q),$$

$$Q_9^{(4q)} = (\bar{s}_L \gamma_\mu \gamma_\nu \gamma_\rho b_L) \sum_q e_q (\bar{q} \gamma^\mu \gamma^\nu \gamma^\rho q), \tag{C.2}$$

$$Q_{10}^{(4q)} = (\bar{s}_L \gamma_\mu \gamma_\nu \gamma_\rho T^a b_L) \sum_q e_q (\bar{q} \gamma^\mu \gamma^\nu \gamma^\rho T^a q),$$

where the sum runs over $q = u, d, s, c, b$. According to the one-loop topologies that generate non-vanishing coefficients for these operators, those in Eq. (C.1) are denoted QCD-penguin operators, whereas those in Eq. (C.2) are referred to as non-leptonic EW-penguin operators.

Both the initial conditions and the anomalous dimensions of all the Wilson coefficients appearing in $\mathcal{L}_{\text{eff}}^{b \to s}$ in Eq. (4.1) are known to high

Table C.1: Numerical values for the Wilson coefficients of the effective four-fermion operators relevant to $b \to s\ell^+\ell^-$ decays, within the SM. The values are reported for different values of the low-energy RG scale $\mu_b$ close to the $b$-quark mass. The last two columns show the uncertainty for $\mu_b = 4.2\,\text{GeV}$ resulting from the variation of the high-energy matching scale $\mu_0$ between 80 and 320 GeV, as well as from the variation of the QCD coupling and the top-quark mass (Table from Ref. [463]).

| | $\mu_b$ [GeV] | | | | $\delta C_i(\mu_b = 4.2\,\text{GeV})$ | |
| | 2.0 | 4.2 | 4.8 | 5.0 | $\mu_0$ | $m_t, \alpha_s$ |
|---|---|---|---|---|---|---|
| $C_1$ | −0.492 | −0.294 | −0.264 | −0.255 | 0.009 | 0.002 |
| $C_2$ | 1.033 | 1.017 | 1.015 | 1.014 | 0.001 | 0.000 |
| $C_3^{(4q)}$ | −0.0133 | −0.0059 | −0.0051 | −0.0048 | 0.0002 | 0.0001 |
| $C_4^{(4q)}$ | −0.147 | −0.087 | −0.080 | −0.078 | 0.000 | 0.001 |
| $C_5^{(4q)}$ | 0.0009 | 0.0004 | 0.0004 | 0.0003 | 0.0000 | 0.0000 |
| $C_6^{(4q)}$ | 0.0030 | 0.0011 | 0.0009 | 0.0009 | 0.0001 | 0.0000 |
| $C_7^{\text{eff}}$ | −0.3189 | −0.2957 | −0.2915 | −0.2902 | 0.0002 | 0.0005 |
| $C_8^{\text{eff}}$ | −0.1780 | −0.1630 | −0.1606 | −0.1599 | 0.0005 | 0.0004 |
| $C_9$ | 4.349 | 4.114 | 4.053 | 4.033 | 0.012 | 0.007 |
| $C_{10}$ | −4.220 | −4.193 | −4.189 | −4.187 | 0.000 | 0.033 |

accuracy, typically at the next-to-next-to-leading order in QCD and at the next-to-leading order in the electroweak interactions (see in particular Refs. [281, 457–462]). For the dipole operators, the effective scheme-independent coefficients $C_{7,8}^{\text{eff}}$ are usually defined

$$C_7^{\text{eff}}(\mu) = C_7(\mu) + \sum_{i=1}^{6} y_i C_i^{(4q)}(\mu), \quad C_8^{\text{eff}}(\mu) = C_8(\mu) + \sum_{i=1}^{6} z_i C_i^{(4q)}(\mu),$$

$$(C.3)$$

where $y = (0, 0, -\frac{1}{3}, -\frac{4}{9}, -\frac{20}{3}, -\frac{80}{9})$ and $z = (0, 0, 1, -\frac{1}{6}, 20, -\frac{10}{3})$, in the naïve dimensional regularisation scheme. Updated numerical values for the Wilson coefficients of the FCNC and the QCD-penguin operators, within the SM, are shown in Table C.1 (from Ref. [463]).

# Appendix D

# Simplified expressions for selected observables

We report here numerical and/or parametrical expressions for a series of $b \to s\ell^+\ell^-$ and $b \to c\tau\bar{\nu}$ observables in terms of Wilson coefficients.

### The LFU ratios $R_K$ and $R_{K^*}$

The precise definition of the LFU ratios $R_K$ and $R_{K^*}$ measured so far is

$$R_{K^{(*)}}^{[q_{\min}^2, q_{\max}^2]} = \frac{\int_{q_{\min}^2}^{q_{\max}^2} dq^2 \frac{d\mathcal{B}}{dq^2}(B \to K^{(*)} \mu^+ \mu^-)}{\int_{q_{\min}^2}^{q_{\max}^2} dq^2 \frac{d\mathcal{B}}{dq^2}(B \to K^{(*)} e^+ e^-)}, \tag{D.1}$$

where the rates are meant to be photon inclusive and, as explicitly indicated, the ratio refers to well-defined dilepton mass intervals. Assuming NP affects only the SM Wilson coefficients $C_9$ and $C_{10}$, as it happens in the models we have considered in this book, the numerical expressions of $R_{K^{(*)}}$ for $1.1\,\mathrm{GeV}^2 < q^2 < 1.6\,\mathrm{GeV}^2$, in terms of the modifications to the $C_i^\ell$ defined in Eq. (4.8), are

$$R_K^{[1.1,6]} = 1 + 0.25\,\mathrm{Re}(\Delta C_9^\mu - \Delta C_{10}^\mu) - 0.01\,\mathrm{Re}(\Delta C_9^\mu + \Delta C_{10}^\mu) + \mathcal{O}(\Delta C_i^2), \tag{D.2}$$

$$R_{K^*}^{[1.1,6]} = 1 + 0.24\,\mathrm{Re}(\Delta C_9^\mu - \Delta C_{10}^\mu) - 0.05\,\mathrm{Re}(\Delta C_9^\mu + \Delta C_{10}^\mu) + \mathcal{O}(\Delta C_i^2). \tag{D.3}$$

As can be seen, in both cases the correction is dominated by left-handed LFU-violating contributions. In the limit of a purely left-handed

LFU-violating amplitude (i.e. for $\Delta C_9^\mu = \Delta C_{10}^\mu$), the NP correction is almost universal.

In the low-$q^2$ bin of $R_{K^*}^{[1.1,6]}$, where the kinematical dependence from the lepton masses is not negligible and where the dipole operator plays a relevant role, one gets

$$R_{K^*}^{[0.045,1.1]} = R_{K^*,\mathrm{SM}}^{[0.045,1.1]} \left[1 + 0.06\,\mathrm{Re}(\Delta C_9^\mu - \Delta C_{10}^\mu) - 0.02\,\mathrm{Re}(\Delta C_9^\mu \right.$$
$$\left. + \Delta C_{10}^\mu) + 0.003\,\mathrm{Re}(\Delta C_9^U) - 0.005\,\mathrm{Re}(\Delta C_{10}^U) + \mathcal{O}(\Delta C_i^2)\right], \tag{D.4}$$

where [187]

$$R_{K^*,\mathrm{SM}}^{[0.045,1.1]} = 0.906 \pm 0.028. \tag{D.5}$$

Since the sizable contribution of the dipole operator is necessarily lepton universal, in this case the NP impact is naturally smaller. Note also that the lepton-universal correction terms $\Delta C_{9,10}^U$ have a small but non-negligible impact. This is because of the non-negligible kinematical dependence from the lepton masses, which changes with possible (universal) deformations of the dilepton spectrum.

## $\mathcal{B}(B_s \to \mu^+\mu^-)$

In the class of models where NP affects only the Wilson coefficients $C_{9,10}$, we can write

$$\mathcal{B}(B_i \to \ell^+\ell^-) = \mathcal{B}(B_i \to \ell^+\ell^-)_{\mathrm{SM}} \left|1 + \frac{\Delta C_{10}^\mu + \Delta C_{10}^U}{C_{10}^{\mathrm{SM}}}\right|, \tag{D.6}$$

with [281–283]

$$\mathcal{B}(B_i \to \ell^+\ell^-)_{\mathrm{SM}} = \frac{4G_F}{\sqrt{2}} \frac{\alpha_{\mathrm{em}}}{4\pi} |V_{tb}^* V_{ts}|^2 \frac{\tau_{B_s} f_{B_s}^2 m_{B_s} m_\mu^2}{8\pi}$$
$$\times \sqrt{1 - \frac{4m_\mu^2}{m_{B_s^0}^2}} |C_{10}|^2 [1 + \delta_{\mathrm{QED}}]$$
$$= (3.66 \pm 0.14) \times 10^{-9}. \tag{D.7}$$

Here $\delta_{\mathrm{QED}}$ collectively denotes the QCD-QED corrections recently estimated in Ref. [213], which amount to a 0.5% reduction of the branching ratio. The error is dominated by parametric uncertainties on $f_{B_s}$ and $|V_{ts}|$ [213].

$R_D$, $R_{D^*}$, and $\mathcal{B}(B^+ \to \tau^+ \nu)$

Using the results in [233, 249] for the $\bar{B} \to D^{(*)} \ell \bar{\nu}$ form factors and decay rates, and employing the definition of the $b \to c \tau \bar{\nu}$ effective Lagrangian in Eq. (4.19), the LFU ratios $R_H = \Gamma(\bar{B} \to H \tau \bar{\nu})/\Gamma(\bar{B} \to H \ell \bar{\nu})$, for $H = D$ and $D^*$, can be expressed as

$$\frac{R_D}{R_D^{\text{SM}}} \approx |1 + C_{V_L}^c|^2 + 1.50(1) \operatorname{Re}[(1 + C_{V_L}^c) \eta_S C_{S_R}^{c*}] + 1.03(1) |\eta_S C_{S_R}^c|^2,$$

$$\frac{R_{D^*}}{R_{D^*}^{\text{SM}}} \approx |1 + C_{V_L}^c|^2 + 0.12(1) \operatorname{Re}[(1 + C_{V_L}^c) \eta_S C_{S_R}^{c*}] + 0.04(1) |\eta_S C_{S_R}^c|^2.$$

$$(D.8)$$

Here $\eta_S$ denotes the effect of the RG evolution of scalar operator from a generic high scale $\Lambda$ down to $m_b$ [464]. Setting $\Lambda = 2$ TeV, leads to $\eta_S \approx 1.7$, which is the value used to derive the numerical coefficients in Eq. (4.20). The updated SM predictions for $R_{D^{(*)}}$ by the HFLAV Collaboration [76] are (see also Refs. [267, 465]):

$$R_D^{\text{SM}} = 0.299 \pm 0.003, \quad R_{D^*}^{\text{SM}} = 0.258 \pm 0.005. \quad (D.9)$$

Concerning $b \to u$ transitions, the branching ration of the dilepton decay $B \to \tau \bar{\nu}$ can be written as

$$\mathcal{B}(\bar{B} \to \tau \bar{\nu}) = \mathcal{B}(\bar{B} \to \tau \bar{\nu})_{\text{SM}} \left| 1 + C_{V_L}^u + C_{S_R}^u \frac{m_B^2}{m_\tau (m_b + m_u)} \right|^2, \quad (D.10)$$

where $m_{b,u}$ denote $\overline{\text{MS}}$ quark masses to be evaluated at the same (high) scale $\Lambda$ at which $C_{S_R}^u$ is evaluated. The SM expression is

$$\mathcal{B}(\bar{B} \to \tau \bar{\nu})_{\text{SM}} = G_F |V_{ub}|^2 \frac{\tau_B f_B^2 m_B m_\tau^2}{8\pi} \left( 1 - \frac{m_\tau^2}{m_B^2} \right)^2$$

$$= (0.812 \pm 0.054) \times 10^{-4}, \quad (D.11)$$

where the error is dominated by the parametric uncertainty on $|V_{ub}|$ [153].

# Bibliography

[1] S. L. Glashow, *Partial symmetries of weak interactions*, Nucl. Phys. **22** (1961) 579.

[2] S. Weinberg, *A model of leptons*, Phys. Rev. Lett. **19** (1967) 1264.

[3] A. Salam, *Weak and electromagnetic interactions*, Conf. Proc. C **680519** (1968) 367.

[4] D. J. Gross and F. Wilczek, *Ultraviolet behavior of nonabelian gauge theories*, Phys. Rev. Lett. **30** (1973) 1343.

[5] H. D. Politzer, *Reliable perturbative results for strong interactions?*, Phys. Rev. Lett. **30** (1973) 1346.

[6] F. Englert and R. Brout, *Broken symmetry and the mass of gauge vector mesons*, Phys. Rev. Lett. **13** (1964) 321; P. W. Higgs, *Broken symmetries and the masses of gauge bosons*, Phys. Rev. Lett. **13** (1964) 508; G. S. Guralnik, C. R. Hagen, and T. W. B. Kibble, *Global conservation laws and massless particles*, Phys. Rev. Lett. **13** (1964) 585.

[7] ATLAS, G. Aad et al., *Observation of a new particle in the search for the Standard Model Higgs boson with the ATLAS detector at the LHC*, Phys. Lett. B **716** (2012) 1, arXiv:1207.7214.

[8] CMS, S. Chatrchyan et al., *Observation of a new boson at a mass of 125 GeV with the CMS experiment at the LHC*, Phys. Lett. B **716** (2012) 30, arXiv:1207.7235.

[9] ATLAS, CMS, G. Aad et al., *Measurements of the Higgs boson production and decay rates and constraints on its couplings from a combined ATLAS and CMS analysis of the LHC pp collision data at $\sqrt{s} = 7$ and 8 TeV*, JHEP **08** (2016) 045, arXiv:1606.02266.

[10] N. Cabibbo, *Unitary symmetry and leptonic decays*, Phys. Rev. Lett. **10** (1963) 531.

[11] M. Kobayashi and T. Maskawa, *CP violation in the renormalizable theory of weak interaction*, Prog. Theor. Phys. **49** (1973) 652.

[12] Particle Data Group, P. A. Zyla et al., *Review of Particle Physics*, PTEP **2020** (2020) 083C01.

[13] CLEO, R. Ammar *et al.*, *Evidence for penguins: First observation of $B \to K^*(892)\gamma$*, Phys. Rev. Lett. **71** (1993) 674.

[14] CLEO, M. S. Alam *et al.*, *First measurement of the rate for the inclusive radiative penguin decay $b \to s\gamma$*, Phys. Rev. Lett. **74** (1995) 2885.

[15] M. Misiak, A. Rehman, and M. Steinhauser, *Towards $\overline{B} \to X_s\gamma$ at the NNLO in QCD without interpolation in $m_c$*, JHEP **06** (2020) 175, arXiv:2002.01548.

[16] ARGUS, H. Albrecht *et al.*, *ARGUS: A universal detector at DORIS-II*, Nucl. Instrum. Meth. A **275** (1989) 1.

[17] CLEO II, Y. Kubota *et al.*, *The CLEO-II detector*, Nucl. Instrum. Meth. A **320** (1992) 66.

[18] CLEO III collaboration, S. E. Kopp, *The CLEO III detector*, Nucl. Instrum. Meth. A **384** (1996) 61.

[19] CDF, F. Abe *et al.*, *The CDF Detector: An Overview*, Nucl. Instrum. Meth. A **271** (1988) 387.

[20] D0, S. Abachi *et al.*, *The D0 Detector*, Nucl. Instrum. Meth. A **338** (1994) 185.

[21] Belle, A. Abashian *et al.*, *The Belle Detector*, Nucl. Instrum. Meth. A **479** (2002) 117.

[22] BaBar collaboration, B. Aubert *et al.*, *The BaBar detector*, Nucl. Instrum. Meth. A **479** (2002) 1, arXiv:hep-ex/0105044.

[23] M. Jung, *Branching ratio measurements and isospin violation in B-meson decays*, Phys. Lett. B **753** (2016) 187, arXiv:1510.03423.

[24] Belle-II, W. Altmannshofer *et al.*, *The Belle II Physics Book*, PTEP **2019** (2019) 123C01, arXiv:1808.10567, [Erratum: PTEP 2020, 029201 (2020)].

[25] CLEO, T. Skwarnicki, *Initial performance of the CLEO-II CsI calorimeter*, Conf. Proc. C **900802V2** (1990) C900802V2:1359.

[26] Belle, R. Louvot, *Y(5S) Results at Belle*, in *24th Lake Louise Winter Institute: Fundamental interactions*, 2009, arXiv:0905.4345.

[27] ATLAS, G. Aad *et al.*, *The ATLAS Experiment at the CERN Large Hadron Collider*, JINST **3** (2008) S08003.

[28] CMS, S. Chatrchyan *et al.*, *The CMS Experiment at the CERN LHC*, JINST **3** (2008) S08004.

[29] LHCb, A. A. Alves Jr. *et al.*, *The LHCb Detector at the LHC*, JINST **3** (2008) S08005.

[30] LHCb, R. Aaij *et al.*, *Measurement of the b-quark production cross-section in 7 and 13 TeV pp collisions*, Phys. Rev. Lett. **118** (2017) 052002, arXiv:1612.05140, [Erratum: Phys.Rev.Lett. 119, 169901 (2017)].

[31] LHCb, S. Akar, *The LHCb upgrade*, J. Phys. Conf. Ser. **556** (2014) 012073.

[32] L.-L. Chau and W.-Y. Keung, *Comments on the Parametrization of the Kobayashi-Maskawa Matrix*, Phys. Rev. Lett. **53** (1984) 1802.

[33] L. Wolfenstein, *Parametrization of the Kobayashi-Maskawa Matrix*, Phys. Rev. Lett. **51** (1983) 1945.

[34] CKMfitter Group, J. Charles *et al.*, *CP violation and the CKM matrix: Assessing the impact of the asymmetric B factories*, Eur. Phys. J. C **41** (2005) 1, arXiv:hep-ph/0406184.

[35] I. I. Bigi and A. I. Sanda, *CP Violation*, vol. 9, Cambridge University Press, 2009.

[36] A. Buras, *Gauge Theory of Weak Decays*, Cambridge University Press, 2020.

[37] N. Isgur and M. B. Wise, *Weak decays of heavy mesons in the static quark approximation*, Phys. Lett. B **232** (1989) 113.

[38] N. Isgur and M. B. Wise, *Weak transition form-factors between heavy mesons*, Phys. Lett. B **237** (1990) 527.

[39] H. Georgi, *An effective field theory for heavy quarks at low-energies*, Phys. Lett. B **240** (1990) 447.

[40] A. V. Manohar and M. B. Wise, *Heavy Quark Physics*, Cambridge University Press, 2000.

[41] T. Mannel, *Effective field theories for heavy quarks: Heavy quark effective theory and Hheavy quark expansion*, Les Houches Lect. Notes **108** (2020).

[42] C. W. Bauer, S. Fleming, D. Pirjol, and I. W. Stewart, *An Effective field theory for collinear and soft gluons: Heavy to light decays*, Phys. Rev. D **63** (2001) 114020, arXiv:hep-ph/0011336.

[43] C. W. Bauer, D. Pirjol, and I. W. Stewart, *Soft collinear factorization in effective field theory*, Phys. Rev. D **65** (2002) 054022, arXiv:hep-ph/0109045.

[44] N. Isgur and M. B. Wise, *Weak transition form-factors between heavy mesons*, Phys. Lett. B **237** (1990) 527.

[45] M. Neubert, *Heavy quark symmetry*, Phys. Rept. **245** (1994) 259, arXiv:hep-ph/9306320.

[46] Flavour Lattice Averaging Group, S. Aoki *et al.*, *FLAG Review 2019: Flavour Lattice Averaging Group (FLAG)*, Eur. Phys. J. C **80** (2020) 113, arXiv:1902.08191.

[47] A. Khodjamirian, R. Ruckl, S. Weinzierl, C. W. Winhart, and O. I. Yakovlev, *Predictions on $B \to \pi \ell \nu$, $D \to \pi \ell \bar{\nu}$, and $D \to K \ell \bar{\nu}$ from QCD light cone sum rules*, Phys. Rev. D **62** (2000) 114002, arXiv:hep-ph/0001297.

[48] P. Ball and R. Zwicky, *New results on $B \to \pi, K, \eta$ decay form-factors from light-cone sum rules*, Phys. Rev. D **71** (2005) 014015, arXiv:hep-ph/0406232.

[49] A. Khodjamirian, T. Mannel, and N. Offen, *Form-factors from light-cone sum rules with B-meson distribution amplitudes*, Phys. Rev. D **75** (2007) 054013, arXiv:hep-ph/0611193.

[50] N. Gubernari, A. Kokulu, and D. van Dyk, *$B \to P$ and $B \to V$ Form Factors from B-Meson Light-Cone Sum Rules beyond Leading Twist*, JHEP **01** (2019) 150, arXiv:1811.00983.

[51] Y. Y. Keum, H.-N. Li, and A. I. Sanda, *Penguin enhancement and $B \to K\pi$ decays in perturbative QCD*, Phys. Rev. D **63** (2001) 054008, arXiv:hep-ph/0004173.

[52] M. Beneke, G. Buchalla, M. Neubert, and C. T. Sachrajda, *QCD factorization for exclusive, nonleptonic B meson decays: General arguments and the case of heavy light final states*, Nucl. Phys. B **591** (2000) 313, arXiv:hep-ph/0006124.

[53]  M. Beneke and M. Neubert, *QCD factorization for B —> PP and B —>
      PV decays*, Nucl. Phys. B **675** (2003) 333, arXiv:hep-ph/0308039.

[54]  A. Ali *et al.*, *Charmless non-leptonic $B_s$ decays to PP, PV and VV
      final states in the pQCD approach*, Phys. Rev. D **76** (2007) 074018,
      arXiv:hep-ph/0703162.

[55]  C. W. Bauer, I. Z. Rothstein, and I. W. Stewart, *SCET analysis of B —>
      K pi, B —> K anti-K, and B —> pi pi decays*, Phys. Rev. D **74** (2006)
      034010, arXiv:hep-ph/0510241.

[56]  S. Weinberg, *Baryon and lepton nonconserving processes*, Phys. Rev. Lett.
      **43** (1979) 1566.

[57]  W. Buchmuller and D. Wyler, *Effective Lagrangian analysis of new
      interactions and flavor conservation*, Nucl. Phys. B **268** (1986) 621.

[58]  B. Grzadkowski, M. Iskrzynski, M. Misiak, and J. Rosiek, *Dimension-
      Six Terms in the Standard Model Lagrangian*, JHEP **10** (2010) 085,
      arXiv:1008.4884.

[59]  E. E. Jenkins, A. V. Manohar, and M. Trott, *Renormalization group
      evolution of the Standard Model dimension six operators I: Formalism and
      λ dependence*, JHEP **10** (2013) 087, arXiv:1308.2627.

[60]  E. E. Jenkins, A. V. Manohar, and M. Trott, *Renormalization group
      evolution of the Standard Model dimension six operators II: Yukawa
      dependence*, JHEP **01** (2014) 035, arXiv:1310.4838.

[61]  R. Alonso, E. E. Jenkins, A. V. Manohar, and M. Trott, *Renormaliza-
      tion group evolution of the Standard Model dimension six operators III:
      Gauge coupling dependence and phenomenology*, JHEP **04** (2014) 159,
      arXiv:1312.2014.

[62]  M. Wirbel, B. Stech, and M. Bauer, *Exclusive Semileptonic Decays of Heavy
      Mesons*, Z. Phys. C **29** (1985) 637.

[63]  N. Isgur, D. Scora, B. Grinstein, and M. B. Wise, *Semileptonic B and D
      Decays in the Quark Model*, Phys. Rev. D **39** (1989) 799.

[64]  A. Sirlin, *Large $m_W$, $m_Z$ behaviour of the $\ell(\alpha)$ corrections to semileptonic
      processes mediated by w*, Nuclear Physics B **196** (1982) 83.

[65]  J. A. Bailey *et al.*, *Update of $|v_{cb}|$ from the → star$\ell\bar\nu$ form factor at zero
      recoil with three-flavor lattice qcd*, Physical Review D **89** (2014).

[66]  I. Caprini, L. Lellouch, and M. Neubert, *Dispersive bounds on the
      shape of $\bar{B} \to D^{(*)}\ell^-\bar\nu$ form-factors*, Nucl. Phys. B **530** (1998) 153,
      arXiv:hep-ph/9712417.

[67]  B. Grinstein and A. Kobach, *Model-independent extraction of $|V_{ub}|$ from
      $B \to d^*\ell n\bar u$*, Physics Letters B **771** (2017) 359–364.

[68]  Flavour Lattice Averaging Group, S. Aoki *et al.*, *FLAG Review 2019:
      Flavour Lattice Averaging Group (FLAG)*, Eur. Phys. J. C **80** (2020) 113,
      arXiv:1902.08191.

[69]  HPQCD, H. Na, C. M. Bouchard, G. P. Lepage, C. Monahan, and
      J. Shigemitsu, *$B \to Dl\nu$ form factors at nonzero recoil and extraction
      of $|V_{cb}|$*, Phys. Rev. D **92** (2015) 054510, arXiv:1505.03925, [Erratum:
      Phys.Rev.D 93, 119906 (2016)].

[70] HPQCD, J. Harrison, C. Davies, and M. Wingate, *Lattice QCD calculation of the $B_{(s)} \to D^*_{(s)} \ell \nu$ form factors at zero recoil and implications for $|V_{cb}|$*, Phys. Rev. D **97** (2018) 054502, arXiv:1711.11013.

[71] LHCb, R. Aaij *et al.*, *Measurement of $|V_{cb}|$ with $B_s^0 \to D_s^{(*)-} \mu^+ \nu_\mu$ decays*, Phys. Rev. **D101** (2020) 072004, arXiv:2001.03225.

[72] LHCb, R. Aaij *et al.*, *Measurement of the shape of the $\Lambda_b^0 \to \Lambda_c^+ \mu^- \overline{\nu}_\mu$ differential decay rate*, Phys. Rev. D **96** (2017) 112005, arXiv:1709.01920.

[73] Fermilab Lattice, MILC, A. Bazavov *et al.*, *Semileptonic form factors for $B \to D^* \ell \nu$ at nonzero recoil from 2 + 1-flavor lattice QCD*, arXiv:2105.14019.

[74] HFLAV collaboration, Y. S. Amhis *et al.*, *Averages of b-hadron, c-hadron, and $\tau$-lepton properties as of 2018*, arXiv:1909.12524, updated results and plots available at https://hflav.web.cern.ch/.

[75] C. G. Boyd, B. Grinstein, and R. F. Lebed, *Precision corrections to dispersive bounds on form-factors*, Phys. Rev. D **56** (1997) 6895, arXiv:hep-ph/9705252.

[76] HFLAV, Y. S. Amhis *et al.*, *Averages of b-hadron, c-hadron, and $\tau$-lepton properties as of 2018*, Eur. Phys. J. C **81** (2021) 226, arXiv:1909.12524.

[77] I. I. Y. Bigi, M. A. Shifman, N. Uraltsev, and A. I. Vainshtein, *High power n of m(b) in beauty widths and $n=5 \to$ infinity limit*, Phys. Rev. D **56** (1997) 4017, arXiv:hep-ph/9704245.

[78] P. Gambino and C. Schwanda, *Inclusive semileptonic fits, heavy quark masses, and $v_{cb}$*, Physical Review D **89** (2014).

[79] A. Alberti, P. Gambino, K. J. Healey, and S. Nandi, *Precision determination of the Cabibbo-Kobayashi-Maskawa element $V_{cb}$*, Phys. Rev. Lett. **114** (2015) 061802, arXiv:1411.6560.

[80] K. G. Chetyrkin *et al.*, *Charm and bottom quark masses: An update*, Phys. Rev. D **80** (2009) 074010, arXiv:0907.2110.

[81] I. I. Y. Bigi, M. A. Shifman, and N. Uraltsev, *Aspects of heavy quark theory*, Ann. Rev. Nucl. Part. Sci. **47** (1997) 591, arXiv:hep-ph/9703290.

[82] M. Bordone, B. Capdevila, and P. Gambino, *Three Loop Calculations and Inclusive $V_{cb}$*, Phys. Lett. B **822** (2021) 136679, arXiv:2107.00604.

[83] A. H. Hoang, Z. Ligeti, and A. V. Manohar, *b decays in the upsilon expansion*, Physical Review D **59** (1999).

[84] N. Uraltsev, *Theoretical uncertainties in $\Gamma_{sl}(b \to u)$*, Int. J. Mod. Phys. A **14** (1999) 4641, arXiv:hep-ph/9905520.

[85] M. Neubert, *Analysis of the photon spectrum in inclusive $b \to X_s \gamma$ decays*, Physical Review D **49** (1994) 4623–4633.

[86] I. I. Bigi, M. A. Shifman, N. G. Uraltsev, and A. L. Vainshtein, *On the motion of heavy quarks inside hadrons: universal distributions and inclusive decays*, International Journal of Modern Physics A **09** (1994) 2467–2504.

[87] B. O. Lange, M. Neubert, and G. Paz, *Theory of charmless inclusive b decays and the extraction of $|V_{ub}|$*, Physical Review D **72** (2005).

[88] P. Gambino, P. Giordano, G. Ossola, and N. Uraltsev, *Inclusive semileptonic B decays and the determination of $|V_{ub}|$*, JHEP **10** (2007) 058.

[89]  E. Gardi, *On the determination of $|V_{ub}|$ from inclusive semileptonic B decays*, Frascati Phys. Ser. **47** (2008) 381, arXiv:0806.4524.

[90]  U. Aglietti, F. Di Lodovico, G. Ferrera, and G. Ricciardi, *Inclusive measure of $|V_{ub}|$ with the analytic coupling model*, Eur. Phys. J. C **59** (2009) 831, arXiv:0711.0860.

[91]  C. W. Bauer, Z. Ligeti, and M. E. Luke, *Precision determination of $|V_{ub}|$ from inclusive decays*, Phys. Rev. D **64** (2001) 113004, arXiv:hep-ph/0107074.

[92]  Belle, L. Cao et al., *Measurements of partial branching fractions of inclusive $B \to X_u \ell + \nu \ell$ decays with hadronic tagging*, Phys. Rev. D **104** (2021) 012008, arXiv:2102.00020.

[93]  J. R. Andersen and E. Gardi, *Inclusive spectra in charmless semileptonic b decays by dressed gluon exponentiation*, Journal of High Energy Physics **2006** (2006) 097–097.

[94]  LHCb, R. Aaij et al., *Determination of the quark coupling strength $|V_{ub}|$ using baryonic decays*, Nature Phys. **11** (2015) 743, arXiv:1504.01568.

[95]  W. Detmold, C. Lehner, and S. Meinel, *$\Lambda_b \to p\ell^- \bar{\nu}_\ell$ and $\Lambda_b \to \Lambda_c \ell^- \bar{\nu}_\ell$ form factors from lattice QCD with relativistic heavy quarks*, Phys. Rev. D **92** (2015) 034503, arXiv:1503.01421.

[96]  CLEO, A. H. Mahmood et al., *Measurement of the B-meson inclusive semileptonic branching fraction and electron energy moments*, Phys. Rev. D **70** (2004) 032003, arXiv:hep-ex/0403053.

[97]  BaBar, B. Aubert et al., *Measurement of the ratio $\mathcal{B}B(B^+ \to Xe\nu)/\mathcal{B}B(B^0 \to Xe\nu)$*, Phys. Rev. D **74** (2006) 091105, arXiv:hep-ex/0607111.

[98]  Belle, P. Urquijo et al., *Moments of the electron energy spectrum and partial branching fraction of $B \to X(c)e\nu$ decays at Belle*, Phys. Rev. D **75** (2007) 032001, arXiv:hep-ex/0610012.

[99]  I. I. Bigi, T. Mannel, and N. Uraltsev, *Semileptonic width ratios among beauty hadrons*, JHEP **09** (2011) 012, arXiv:1105.4574.

[100]  M. Rudolph, *An experimentalist's guide to the semileptonic bottom to charm branching fractions*, Int. J. Mod. Phys. A **33** (2018) 1850176, arXiv:1805.05659.

[101]  F. U. Bernlochner, Z. Ligeti, and S. Turczyk, *A Proposal to solve some puzzles in semileptonic B decays*, Phys. Rev. D **85** (2012) 094033, arXiv:1202.1834.

[102]  J. Brod and J. Zupan, *The ultimate theoretical error on $\gamma$ from $B \to DK$ decays*, JHEP **01** (2014) 051, arXiv:1308.5663.

[103]  M. Gronau and D. London, *How to determine all the angles of the unitarity triangle from $B^0 \to DK_s$ and $B_s^0 \to D\phi$*, Phys. Lett. B **253** (1991) 483.

[104]  M. Gronau and D. Wyler, *On determining a weak phase from CP asymmetries in charged B decays*, Phys. Lett. B **265** (1991) 172.

[105]  D. Atwood, I. Dunietz, and A. Soni, *Enhanced CP violation with $B \to KD^0$ $(\overline{D}^0)$ modes and extraction of the CKM angle $\gamma$*, Phys. Rev. Lett. **78** (1997) 3257, arXiv:hep-ph/9612433.

[106] A. Giri, Y. Grossman, A. Soffer, and J. Zupan, *Determining $\gamma$ using $B^{\pm} \to DK^{\pm}$ with multibody D decays*, Phys. Rev. D **68** (2003) 054018, arXiv:hep-ph/0303187.

[107] Belle, A. Poluektov *et al.*, *Measurement of $\phi(3)$ with Dalitz plot analysis of $B^{\pm} \to D^{**(*)}K^{\pm}$ decay*, Phys. Rev. D **70** (2004) 072003, arXiv:hep-ex/0406067.

[108] Y. Grossman, Z. Ligeti, and A. Soffer, *Measuring $\gamma$ in $B^{\pm} \to K^{\pm}(KK^*)(D)$ decays*, Phys. Rev. D **67** (2003) 071301, arXiv:hep-ph/0210433.

[109] R. Aleksan, I. Dunietz, and B. Kayser, *Determining the CP violating phase $\gamma$*, Z. Phys. C **54** (1992) 653.

[110] R. Fleischer, *New strategies to obtain insights into CP violation through $B_s^0 \to D_s^{\pm}K^{\mp}$, $D_s^{*\pm}K^{\mp}$, ... and $B_d^0 \to D^{\pm}\pi^{\mp}$, $D^{*\pm}\pi^{\mp}$, ... decays*, Nucl. Phys. B **671** (2003) 459, arXiv:hep-ph/0304027.

[111] LHCb, R. Aaij *et al.*, *Measurement of the CKM angle $\gamma$ and $B_s^0$-$\bar{B}_s^0$ mixing frequency with $B_s^0 \to D_s^{\mp}h^{\pm}\pi^{\pm}\pi^{\mp}$ decays*, JHEP **03** (2021) 137, arXiv:2011.12041.

[112] LHCb Collaboration, *Updated LHCb combination of the CKM angle $\gamma$*, tech. rep., CERN, Geneva, 2020.

[113] LHCb, R. Aaij *et al.*, *Measurement of the CKM angle $\gamma$ from a combination of LHCb results*, JHEP **12** (2016) 087, arXiv:1611.03076.

[114] UT fit collaboration. http://www.utfit.org/UTfit/.

[115] CKMFitter group. http://ckmfitter.in2p3.fr/www/html/ckm_main.html.

[116] M. Artuso, G. Borissov, and A. Lenz, *CP violation in the $B_s^0$ system*, Rev. Mod. Phys. **88** (2016) 045002, arXiv:1511.09466, [Addendum: Rev.Mod.Phys. 91, 049901 (2019)].

[117] Belle, K. Abe *et al.*, *Improved measurement of CP-violation parameters $\sin(2\phi(1))$ and $|\lambda|$, B meson lifetimes, and $B^0 - \bar{B}^0$ mixing parameter $\Delta m(d)$*, Phys. Rev. D **71** (2005) 072003, arXiv:hep-ex/0408111, [Erratum: Phys.Rev.D 71, 079903 (2005)].

[118] LHCb, R. Aaij *et al.*, *Precise determination of the $B_s^0 - \bar{B}_s^0$ oscillation frequency*, Nature Phys. **18** (2022) 1, arXiv:2104.04421.

[119] CDF, A. Abulencia *et al.*, *Observation of $B_s^0 - \bar{B}_s^0$ oscillations*, Phys. Rev. Lett. **97** (2006) 242003, arXiv:hep-ex/0609040.

[120] LHCb, R. Aaij *et al.*, *Precision measurement of the $B_s^0 - \bar{B}_s^0$ oscillation frequency with the decay $B_s^0 \to D_s^-\pi^+$*, New J. Phys. **15** (2013) 053021, arXiv:1304.4741.

[121] LHCb, R. Aaij *et al.*, *Precise determination of the $B^0 - \bar{B}^0$ oscillation frequency*, arXiv:2104.04421.

[122] CLEO, B. H. Behrens *et al.*, *Precise measurement of $B^0\bar{B}^0$ mixing parameters at the $\Upsilon(4S)$*, Phys. Lett. B **490** (2000) 36, arXiv:hep-ex/0005013.

[123] CLEO, D. E. Jaffe *et al.*, *Bounds on the CP asymmetry in like sign dileptons from $B^0\bar{B}^0$ meson decays*, Phys. Rev. Lett. **86** (2001) 5000, arXiv:hep-ex/0101006.

[124]  BaBar, B. Aubert *et al.*, *Limits on the decay-rate difference of neutral B mesons and on CP, T, and CPT violation in $B^0\overline{B}^0$ oscillations*, Phys. Rev. Lett. **92** (2004) 181801, arXiv:hep-ex/0311037.

[125]  BaBar, J. P. Lees *et al.*, *Search for CP Violation in $B^0 - \bar{B}^0$ mixing using partial reconstruction of $B^0 \to D^{*-}X\ell^+\nu_\ell$ and a kaon tag*, Phys. Rev. Lett. **111** (2013) 101802, arXiv:1305.1575, [Addendum: Phys. Rev. Lett. 111, 159901 (2013)].

[126]  BaBar, J. P. Lees *et al.*, *Study of CP asymmetry in $B^0 - \bar{B}^0$ mixing with inclusive dilepton events*, Phys. Rev. Lett. **114** (2015) 081801, arXiv:1411.1842.

[127]  D0, V. M. Abazov *et al.*, *Study of CP-violating charge asymmetries of single muons and like-sign dimuons in $p\bar{p}$ collisions*, Phys. Rev. D **89** (2014) 012002, arXiv:1310.0447.

[128]  LHCb, R. Aaij *et al.*, *Measurement of the semileptonic CP asymmetry in $B^0 - \bar{B}^0$ mixing*, Phys. Rev. Lett. **114** (2015) 041601, arXiv:1409.8586.

[129]  OPAL, K. Ackerstaff *et al.*, *A study of B meson oscillations using hadronic $Z^0$ decays containing leptons*, Z. Phys. C **76** (1997) 401, arXiv:hep-ex/9707009.

[130]  ALEPH, R. Barate *et al.*, *Investigation of inclusive CP asymmetries in $B^0$ decays*, Eur. Phys. J. C **20** (2001) 431.

[131]  BaBar, B. Aubert *et al.*, *Search for T, CP and CPT violation in $B^0 \bar{B}^0$ mixing with inclusive dilepton events*, Phys. Rev. Lett. **96** (2006) 251802, arXiv:hep-ex/0603053.

[132]  Belle, E. Nakano *et al.*, *Charge asymmetry of same-sign dileptons in $B^0 - \bar{B}^0$ mixing*, Phys. Rev. D **73** (2006) 112002, arXiv:hep-ex/0505017.

[133]  D0, V. M. Abazov *et al.*, *Measurement of the semileptonic charge asymmetry using $B_s^0 \to D_s\mu X$ Decays*, Phys. Rev. Lett. **110** (2013) 011801, arXiv:1207.1769.

[134]  LHCb, R. Aaij *et al.*, *Measurement of the CP asymmetry in $B_s^0 - \overline{B}_s^0$ mixing*, Phys. Rev. Lett. **117** (2016) 061803, arXiv:1605.09768, [Addendum: Phys.Rev.Lett. 118, 129903 (2017)].

[135]  S. Stone and L. Zhang, *S-waves and the measurement of CP violating phases in $B_s$ decays*, Phys. Rev. D **79** (2009) 074024, arXiv:0812.2832.

[136]  L. Zhang and S. Stone, *Time-dependent Dalitz-plot formalism for $B_q \to J/\psi\, h^+h^-$*, Phys. Lett. B **719** (2013) 383, arXiv:1212.6434.

[137]  LHCb, R. Aaij *et al.*, *Updated measurement of time-dependent CP-violating observables in $B_s^0 \to J/\psi K^+K^-$ decays*, Eur. Phys. J. C **79** (2019) 706, arXiv:1906.08356, [Erratum: Eur.Phys.J.C 80, 601 (2020)].

[138]  LHCb, R. Aaij *et al.*, *Measurement of resonant and CP components in $\bar{B}_s^0 \to J/\psi\pi^+\pi^-$ decays*, Phys. Rev. **D89** (2014) 092006, arXiv:1402.6248.

[139]  D0, V. M. Abazov *et al.*, *Measurement of the CP-violating phase $\phi_s^{J/\psi\phi}$ using the flavor-tagged decay $B_s^0 \to J/\psi\phi$ in 8 $fb^{-1}$ of $p\bar{p}$ collisions*, Phys. Rev. D **85** (2012) 032006, arXiv:1109.3166.

[140] CDF, T. Aaltonen *et al.*, *Measurement of the bottom-strange meson mixing phase in the full CDF data set*, Phys. Rev. Lett. **109** (2012) 171802, arXiv:1208.2967.

[141] ATLAS, G. Aad *et al.*, *Flavor tagged time-dependent angular analysis of the $B_s \to J/\psi\phi$ decay and extraction of $\Delta\Gamma s$ and the weak phase $\phi_s$ in ATLAS*, Phys. Rev. D **90** (2014) 052007, arXiv:1407.1796.

[142] ATLAS, *Measurement of the CP violation phase $\phi_s$ in $B_s \to J/\psi\phi$ decays in ATLAS at 13 TeV*,

[143] CMS, V. Khachatryan *et al.*, *Measurement of the CP-violating weak phase $\phi_s$ and the decay width difference $\Delta\Gamma_s$ using the $B_s^0 \to J/\psi\phi(1020)$ decay channel in pp collisions at $\sqrt{s} = 8$ TeV*, Phys. Lett. B **757** (2016) 97, arXiv:1507.07527.

[144] LHCb collaboration, R. Aaij *et al.*, *Precision measurement of CP violation in $B_s^0 \to J/\psi K^+ K^-$ decays*, Phys. Rev. Lett. **114** (2015) 041801, arXiv:1411.3104.

[145] LHCb, R. Aaij *et al.*, *First study of the CP -violating phase and decay-width difference in $B_s^0 \to \psi(2S)\phi$ decays*, Phys. Lett. B **762** (2016) 253, arXiv:1608.04855.

[146] LHCb, R. Aaij *et al.*, *Measurement of the CP-violating phase $\phi_s$ in $\bar{B}_s^0 \to D_s^+ D_s^-$ decays*, Phys. Rev. Lett. **113** (2014) 211801, arXiv:1409.4619.

[147] LHCb, R. Aaij *et al.*, *Resonances and CP violation in $B_s^0$ and $\overline{B}_s^0 \to J/\psi K^+ K^-$ decays in the mass region above the $\phi(1020)$*, JHEP **08** (2017) 037, arXiv:1704.08217.

[148] S. L. Glashow, J. Iliopoulos, and L. Maiani, *Weak interactions with lepton-hadron symmetry*, Phys. Rev. D **2** (1970) 1285.

[149] T. Inami and C. S. Lim, *Effects of superheavy quarks and leptons in low-energy weak processes $K_L \to \mu\bar{\mu}$, $K^+ \to \pi^+\nu\bar{\nu}$, and $K^0 \leftrightarrow \overline{K}^0$*, Prog. Theor. Phys. **65** (1981) 297, [Erratum: Prog.Theor.Phys. 65, 1772 (1981)].

[150] A. Badin, F. Gabbiani, and A. A. Petrov, *Lifetime difference in $B_s$ mixing: Standard model and beyond*, Phys. Lett. B **653** (2007) 230, arXiv:0707.0294.

[151] A. Lenz and G. Tetlalmatzi-Xolocotzi, *Model-independent bounds on new physics effects in non-leptonic tree-level decays of b-mesons*, Journal of High Energy Physics **2020** (2020).

[152] G. Isidori, *Flavor physics and CP violation*, in *2012 European School of High-Energy Physics*, 2013, arXiv:1302.0661.

[153] C. Alpigiani *et al.*, *Unitarity Triangle Analysis in the Standard Model and Beyond*, in *5th Large Hadron Collider Physics Conference*, 2017, arXiv:1710.09644.

[154] UTfit, M. Bona *et al.*, *Model-independent constraints on $\Delta F = 2$ operators and the scale of new physics*, JHEP **03** (2008) 049, arXiv:0707.0636.

[155] G. Isidori, Y. Nir, and G. Perez, *Flavor Physics Constraints for Physics Beyond the Standard Model*, Ann. Rev. Nucl. Part. Sci. **60** (2010) 355, arXiv:1002.0900.

[156] R. S. Chivukula, H. Georgi, and L. Randall, *A composite technicolor Standard Model of quarks*, Nucl. Phys. B **292** (1987) 93.

[157] G. D'Ambrosio, G. F. Giudice, G. Isidori, and A. Strumia, *Minimal flavor violation: An effective field theory approach*, Nucl. Phys. B **645** (2002) 155, arXiv:hep-ph/0207036.

[158] LHCb, R. Aaij et al., *Observation of a resonance in $B^+ \to K^+\mu^+\mu^-$ decays at low recoil*, Phys. Rev. Lett. **111** (2013) 112003, arXiv:1307.7595.

[159] Belle, J.-T. Wei et al., *Measurement of the differential branching fraction and forward-backward asymmetry for $B \to K^{(*)}\ell^+\ell^-$*, Phys. Rev. Lett. **103** (2009) 171801, arXiv:0904.0770.

[160] CDF, T. Aaltonen et al., *Measurement of the forward-backward asymmetry in the $B \to K^{(*)}\mu^+\mu^-$ decay and first observation of the $B_s^0 \to \phi\mu^+\mu^-$ decay*, Phys. Rev. Lett. **106** (2011) 161801, arXiv:1101.1028.

[161] BaBar, J. P. Lees et al., *Measurement of branching fractions and rate asymmetries in the rare decays $B \to K^{(*)}\ell^+\ell^-$*, Phys. Rev. D **86** (2012) 032012, arXiv:1204.3933.

[162] CDF, T. Aaltonen et al., *Observation of the baryonic flavor-changing neutral current decay $\Lambda_b^0 \to \Lambda\mu^+\mu^-$*, Phys. Rev. Lett. **107** (2011) 201802, arXiv:1107.3753.

[163] LHCb, R. Aaij et al., *Differential branching fractions and isospin asymmetries of $B \to K^{(*)}\mu^+\mu^-$ decays*, JHEP **06** (2014) 133, arXiv:1403.8044.

[164] C. Bobeth, G. Hiller, D. van Dyk, and C. Wacker, *The decay $B \to K\ell^+\ell^-$ at low hadronic recoil and model-Independent $\Delta B = 1$ constraints*, JHEP **01** (2012) 107, arXiv:1111.2558.

[165] C. Bobeth, G. Hiller, and D. van Dyk, *More benefits of semileptonic rare B decays at low recoil: CP Violation*, JHEP **07** (2011) 067, arXiv:1105.0376.

[166] W. Altmannshofer and D. M. Straub, *New physics in $b \to s$ transitions after LHC run 1*, Eur. Phys. J. C **75** (2015) 382, arXiv:1411.3161.

[167] A. Bharucha, D. M. Straub, and R. Zwicky, *$B \to V\ell^+\ell^-$ in the Standard Model from light-cone sum rules*, JHEP **08** (2016) 098, arXiv:1503.05534.

[168] LHCb, R. Aaij et al., *Differential branching fraction and angular analysis of $\Lambda_b^0 \to \Lambda\mu^+\mu^-$ decays*, JHEP **06** (2015) 115, arXiv:1503.07138, [Erratum: JHEP 09, 145 (2018)].

[169] W. Detmold, C.-J. D. Lin, S. Meinel, and M. Wingate, *$\Lambda_b^0 \to \Lambda\ell^+\ell^-$ form factors and differential branching fraction from lattice QCD*, Phys. Rev. D **87** (2013) 074502, arXiv:1212.4827.

[170] LHCb, R. Aaij et al., *Branching fraction measurements of the rare $B_s^0 \to \phi\mu^+\mu^-$ and $B_s^0 \to f_2'(1525)\mu^+\mu^-$ decays*, Phys. Rev. Lett. **127** (2021) 151801, arXiv:2105.14007.

[171] LHCb, R. Aaij et al., *Angular analysis and differential branching fraction of the decay $B_s^0 \to \phi\mu^+\mu^-$*, JHEP **09** (2015) 179, arXiv:1506.08777.

[172] D. M. Straub, *Flavio: A Python Package for Flavour and Precision Phenomenology in the Standard Model and Beyond*, arXiv:1810.08132.

[173] R. R. Horgan, Z. Liu, S. Meinel, and M. Wingate, *Calculation of $B^0 \to K^{*0}\mu^+\mu^-$ and $B_s^0 \to \phi\mu^+\mu^-$ observables using form factors from lattice QCD*, Phys. Rev. Lett. **112** (2014) 212003, arXiv:1310.3887.

[174] R. R. Horgan, Z. Liu, S. Meinel, and M. Wingate, *Rare B decays using lattice QCD form factors*, PoS **LATTICE2014** (2015) 372, arXiv:1501.00367.

[175] LHCb, R. Aaij *et al.*, *Branching fraction measurements of the rare $B_s^0 \to \phi\mu^+\mu^-$ and $B_s^0 \to f_2'(1525)\mu^+\mu^-$ decays*, arXiv:2105.14007.

[176] G. Hiller and F. Kruger, *More model-independent analysis of $b \to s$ processes*, Phys. Rev. D **69** (2004) 074020, arXiv:hep-ph/0310219.

[177] T. D. Lee and C.-N. Yang, *Question of parity conservation in weak interactions*, Phys. Rev. **104** (1956) 254.

[178] LHCb collaboration, R. Aaij *et al.*, *Test of lepton universality using $B^+ \to K^+\ell^+\ell^-$ decays*, Phys. Rev. Lett. **113** (2014) 151601, arXiv:1406.6482.

[179] LHCb, R. Aaij *et al.*, *Search for lepton-universality violation in $B^+ \to K^+\ell^+\ell^-$ decays*, Phys. Rev. Lett. **122** (2019) 191801, arXiv:1903.09252.

[180] LHCb, R. Aaij *et al.*, *Test of lepton universality in beauty-quark decays*, arXiv:2103.11769.

[181] S. Tolk, J. Albrecht, F. Dettori, and A. Pellegrino, *Data driven trigger efficiency determination at LHCb*, Tech. Rep. LHCb-PUB-2014-039. CERN-LHCb-PUB-2014-039, CERN, Geneva, 2014.

[182] Belle, A. Abdesselam *et al.*, *Test of lepton flavor universality and search for lepton flavor violation in $B \to K\ell\ell$ decays*, JHEP **03** (2021) 105, arXiv:1908.01848.

[183] Belle, S. Wehle *et al.*, *Lepton-flavor-dependent angular analysis of $B \to K^*\ell^+\ell^-$*, Phys. Rev. Lett. **118** (2017) 111801, arXiv:1612.05014.

[184] LHCb collaboration, R. Aaij *et al.*, *Test of lepton universality with $B^0 \to K^{*0}\ell^+\ell^-$ decays*, JHEP **08** (2017) 055, arXiv:1705.05802.

[185] W. D. Hulsbergen, *Decay chain fitting with a Kalman filter*, Nucl. Instrum. Meth. A **552** (2005) 566, arXiv:physics/0503191.

[186] Belle, A. Abdesselam *et al.*, *Test of lepton flavor universality in $B \to K^*\ell^+\ell^-$ decays at Belle*, arXiv:1904.02440.

[187] M. Bordone, G. Isidori, and A. Pattori, *On the Standard Model predictions for $R_K$ and $R_{K^*}$*, Eur. Phys. J. C **76** (2016) 440, arXiv:1605.07633.

[188] S. Descotes-Genon, L. Hofer, J. Matias, and J. Virto, *Global analysis of $b \to s\ell\ell$ anomalies*, JHEP **06** (2016) 092, arXiv:1510.04239.

[189] B. Capdevila, S. Descotes-Genon, J. Matias, and J. Virto, *Assessing lepton-flavour non-universality from $B \to K^*\ell\ell$ angular analyses*, JHEP **10** (2016) 075, arXiv:1605.03156.

[190] B. Capdevila, S. Descotes-Genon, L. Hofer, and J. Matias, *Hadronic uncertainties in $B \to K^*\mu^+\mu^-$: A state-of-the-art analysis*, JHEP **04** (2017) 016, arXiv:1701.08672.

[191] N. Serra, R. Silva Coutinho, and D. van Dyk, *Measuring the breaking of lepton flavor universality in $B \to K^*\ell^+\ell^-$*, Phys. Rev. D **95** (2017) 035029, arXiv:1610.08761.

[192] W. Altmannshofer, C. Niehoff, P. Stangl, and D. M. Straub, *Status of the $B \to K^*\mu^+\mu^-$ anomaly after Moriond 2017*, Eur. Phys. J. C **77** (2017) 377, arXiv:1703.09189.

[193]  S. Jäger and J. Martin Camalich, *Reassessing the discovery potential of the B → K*ℓ⁺ℓ⁻ decays in the large-recoil region: SM challenges and BSM opportunities*, Phys. Rev. D **93** (2016) 014028, arXiv:1412.3183.

[194]  LHCb, R. Aaij *et al.*, *Test of lepton universality with $\Lambda_b^0 \to pK^-\ell^+\ell^-$ decays*, JHEP **05** (2020) 040, arXiv:1912.08139.

[195]  B. Capdevila, A. Crivellin, S. Descotes-Genon, J. Matias, and J. Virto, *Patterns of new physics in $b \to s\ell^+\ell^-$ transitions in the light of recent data*, JHEP **01** (2018) 093, arXiv:1704.05340.

[196]  ATLAS, M. Aaboud *et al.*, *Angular analysis of $B_d^0 \to K^*\mu^+\mu^-$ decays in pp collisions at $\sqrt{s} = 8$ TeV with the ATLAS detector*, JHEP **10** (2018) 047, arXiv:1805.04000.

[197]  CMS, A. M. Sirunyan *et al.*, *Measurement of angular parameters from the decay $B^0 \to K^{*0}\mu^+\mu^-$ in proton-proton collisions at $\sqrt{s} = 8$ TeV*, Phys. Lett. B **781** (2018) 517, arXiv:1710.02846.

[198]  LHCb collaboration, R. Aaij *et al.*, *Measurement of CP-averaged observables in the $B^0 \to K^{*0}\mu^+\mu^-$ decay*, arXiv:2003.04831.

[199]  LHCb, R. Aaij *et al.*, *Measurement of CP-averaged observables in the $B^+ \to K^{*+}\mu^+\mu^-$ decay*, 2021.

[200]  U. Egede, T. Hurth, J. Matias, M. Ramon, and W. Reece, *New observables in the decay mode $\overline{B}_d \to \overline{K}^{*0}\ell^+\ell^-$*, JHEP **11** (2008) 032, arXiv:0807.2589.

[201]  S. Descotes-Genon, J. Matias, M. Ramon, and J. Virto, *Implications from clean observables for the binned analysis of $B \to K^*\mu^+\mu^-$ at large recoil*, JHEP **01** (2013) 048, arXiv:1207.2753.

[202]  C. Bobeth, G. Hiller, and G. Piranishvili, *CP Asymmetries in $\overline{B} \to \overline{K}^*(\to \overline{K}\pi)\overline{\ell}\ell$ and untagged $\overline{B}_s$, $B_s \to \phi(\to K^+K^-)\overline{\ell}\ell$ Decays at NLO*, JHEP **07** (2008) 106, arXiv:0805.2525.

[203]  W. Altmannshofer *et al.*, *Symmetries and Asymmetries of $B \to K^*\mu^+\mu^-$ decays in the Standard Model and beyond*, JHEP **01** (2009) 019, arXiv:0811.1214.

[204]  LHCb, R. Aaij *et al.*, *Differential branching fraction and angular analysis of the decay $B^0 \to K^{*0}\mu^+\mu^-$*, JHEP **08** (2013) 131, arXiv:1304.6325.

[205]  LHCb, R. Aaij *et al.*, *Angular analysis of the $B^0 \to K^{*0}\mu^+\mu^-$ decay using 3 fb⁻¹ of integrated luminosity*, JHEP **02** (2016) 104, arXiv:1512.04442.

[206]  F. Kruger and J. Matias, *Probing new physics via the transverse amplitudes of $B^0 \to K^{*0}(\to K^-\pi^+)l^+l^-$ at large recoil*, Phys. Rev. D **71** (2005) 094009, arXiv:hep-ph/0502060.

[207]  LHCb, R. Aaij *et al.*, *Measurement of the polarization amplitudes in $B^0 \to J/\psi K^*(892)^0$ decays*, Phys. Rev. D **88** (2013) 052002, arXiv:1307.2782; BaBar, B. Aubert *et al.*, *Measurement of decay amplitudes of $B \to J/\psi K^*, \psi(2S)K^*$, and $\chi_{c1}K^*$ with an angular analysis*, Phys. Rev. D **76** (2007) 031102, arXiv:0704.0522.

[208]  A. Khodjamirian, T. Mannel, A. A. Pivovarov, and Y.-M. Wang, *Charm-loop effect in $B \to K^{(*)}\ell^+\ell^-$ and $B \to K^*\gamma$*, JHEP **09** (2010) 089, arXiv:1006.4945.

[209] S. Jäger and J. Martin Camalich, *On B → Vℓℓ at small dilepton invariant mass, power corrections, and new physics*, JHEP **05** (2013) 043, arXiv:1212.2263.

[210] T. Skwarnicki, *A study of the radiative cascade transitions between the Υ' and Υ resonances*, PhD thesis, Cracow, INP, 1986.

[211] ARGUS collaboration, H. Albrecht *et al.*, *Measurement of the polarization in the decay B → J/ψK\**, Phys. Lett. B **340** (1994) 217.

[212] R. Aaij *et al.*, *Angular analysis of the rare decay $B_s^0 → \phi\mu^+\mu^-$*, JHEP **11** (2021) 043, arXiv:2107.13428.

[213] M. Beneke, C. Bobeth, and R. Szafron, *Power-enhanced leading-logarithmic QED corrections to $B_q → \mu^+\mu^-$*, JHEP **10** (2019) 232, arXiv:1908.07011.

[214] K. De Bruyn *et al.*, *Branching ratio measurements of $B_s$ decays*, Phys. Rev. D **86** (2012) 014027, arXiv:1204.1735.

[215] ATLAS, M. Aaboud *et al.*, *Study of the rare decays of $B_s^0$ and $B^0$ mesons into muon pairs using data collected during 2015 and 2016 with the ATLAS detector*, JHEP **04** (2019) 098, arXiv:1812.03017.

[216] CMS, A. M. Sirunyan *et al.*, *Measurement of properties of $B_s^0 → \mu^+\mu^-$ decays and search for $B^0 → \mu^+\mu^-$ with the CMS experiment*, JHEP **04** (2020) 188, arXiv:1910.12127.

[217] LHCb, R. Aaij *et al.*, *Measurement of the $B_s^0 → \mu^+\mu^-$ branching fraction and effective lifetime and search for $B^0 → \mu^+\mu^-$ decays*, Phys. Rev. Lett. **118** (2017) 191801, arXiv:1703.05747.

[218] ATLAS, M. Aaboud *et al.*, *Study of the rare decays of $B_s^0$ and $B^0$ into muon pairs from data collected during the LHC Run 1 with the ATLAS detector*, Eur. Phys. J. C **76** (2016) 513, arXiv:1604.04263.

[219] LHCb, R. Aaij *et al.*, *Measurement of the fragmentation fraction ratio $f_s/f_d$ and its dependence on B meson kinematics*, JHEP **04** (2013) 001, arXiv:1301.5286; LHCb, B. Storaci *et al.*, *Updated average $f_s/f_d$ b-hadron production fraction ratio for 7 TeV pp collisions*.

[220] LHCb, S. Ferreres-Solé, *The beauty of the rare: $B_{(s)} → \mu^+\mu^-$ at the LHCb*, in *55th Rencontres de Moriond on Electroweak Interactions and Unified Theories*, 2021, arXiv:2106.15995.

[221] LHCb, R. Aaij *et al.*, *Precise measurement of the $f_s/f_d$ ratio of fragmentation fractions and of $B_s^0$ decay branching fractions*, Phys. Rev. D **104** (2021) 032005, arXiv:2103.06810.

[222] LHCb, C. Parkes *et al.*, *Improved measurement of $B_{s,d} → \mu^+\mu^-$ decays*, Presented at 9th Edition of the Large Hadron Collider Physics Conference, to appear as LHCb-PAPER-2021-007.

[223] ATLAS, CMS, and LHCbs, *Combination of the ATLAS, CMS and LHCb results on the $B_{(s)}^0 → \mu^+\mu^-$ decays*, https://cds.cern.ch/record/2727207/files/LHCb-CONF-2020-002.pdf, 2020.

[224] Belle, E. Waheed *et al.*, *Measurement of the CKM matrix element $|V_{cb}|$ from $B^0 → D^{*-}\ell^+\nu$ at Belle*, Phys. Rev. D **100** (2019) 052007, arXiv:1809.03290.

[225] Belle, R. Glattauer *et al.*, *Measurement of the decay $B → D\ell\nu_\ell$ in fully reconstructed events and determination of the Cabibbo-Kobayashi-Maskawa matrix element $|V_{cb}|$*, Phys. Rev. D **93** (2016) 032006, arXiv:1510.03657.

[226] BaBar, J. P. Lees *et al.*, *Evidence for an excess of $\bar{B} \to D^{(*)}\tau^-\bar{\nu}_\tau$ decays*, Phys. Rev. Lett. **109** (2012) 101802, arXiv:1205.5442.

[227] BaBar, J. P. Lees *et al.*, *Measurement of an excess of $\bar{B} \to D^{(*)}\tau^-\bar{\nu}_\tau$ decays and implications for charged Higgs bosons*, Phys. Rev. D **88** (2013) 072012, arXiv:1303.0571.

[228] ARGUS, H. Albrecht *et al.*, *Measurement of the decay $B^0 \to D^{*-}\ell^+\nu$*, Phys. Lett. B **197** (1987) 452.

[229] CLEO, D. Bortoletto *et al.*, *Study of the decay $\bar{B}^0 \to D^{*+}\ell^-\bar{\nu}$*, Phys. Rev. Lett. **63** (1989) 1667.

[230] S. Stone, *Semileptonic B decays in book: B decays revised 2nd edition*, 1994. World Scientific Publishing Co. page 283, doi: https://doi.org/10.1142/2425.

[231] M. Feindt *et al.*, *A hierarchical NeuroBayes-based algorithm for full reconstruction of B mesons at B Factories*, Nucl. Instrum. Meth. A **654** (2011) 432, arXiv:1102.3876.

[232] BaBar, B. Aubert *et al.*, *Measurement of $|V_{cb}|$ and the Form-Factor Slope in $\bar{B} \to D\ell^-\bar{\nu}$ decays in events tagged by a fully reconstructed B meson*, Phys. Rev. Lett. **104** (2010) 011802, arXiv:0904.4063.

[233] F. U. Bernlochner, Z. Ligeti, M. Papucci, and D. J. Robinson, *Combined analysis of semileptonic B decays to D and $D^*$: $R(D^{(*)})$, $|V_{cb}|$, and new physics*, Phys. Rev. D **95** (2017) 115008, arXiv:1703.05330, [Erratum: Phys.Rev.D 97, 059902 (2018)].

[234] D. Bigi and P. Gambino, *Revisiting $B \to D\ell\nu$*, Phys. Rev. D **94** (2016) 094008, arXiv:1606.08030.

[235] S. Jaiswal, S. Nandi, and S. K. Patra, *Extraction of $|V_{cb}|$ from $B \to D^{(*)}\ell\nu_\ell$ and the Standard Model predictions of $R(D^{(*)})$*, JHEP **12** (2017) 060, arXiv:1707.09977.

[236] D. Bigi, P. Gambino, and S. Schacht, *$R(D^*)$, $|V_{cb}|$, and the Heavy Quark Symmetry relations between form factors*, JHEP **11** (2017) 061, arXiv:1707.09509.

[237] M. Bordone, N. Gubernari, D. van Dyk, and M. Jung, *Heavy-quark expansion for $\bar{B}_s \to D_s^{(*)}$ form factors and unitarity bounds beyond the $SU(3)_F$ limit*, Eur. Phys. J. C **80** (2020) 347, arXiv:1912.09335.

[238] S. Groote *et al.*, *Form factor independent test of lepton universality in semileptonic heavy meson and baryon decays*, Phys. Rev. D **103** (2021) 093001, arXiv:2102.12818.

[239] Belle, M. Huschle *et al.*, *Measurement of the branching ratio of $\bar{B} \to D^{(*)}\tau^-\bar{\nu}_\tau$ relative to $\bar{B} \to D^{(*)}\ell^-\bar{\nu}_\ell$ decays with hadronic tagging at Belle*, Phys. Rev. D **92** (2015) 072014, arXiv:1507.03233.

[240] Belle, A. Bozek *et al.*, *Observation of $B^+ \to \bar{D}^{*0}\tau^+\nu_\tau$ and Evidence for $B^+ \to \bar{D}^0\tau^+\nu_\tau$ at Belle*, Phys. Rev. D **82** (2010) 072005, arXiv:1005.2302; Belle, I. Adachi *et al.*, *Measurement of B —> D(*) tau nu using full reconstruction tags*, in *24th International Symposium on Lepton-Photon Interactions at High Energy (LP09)*, 2009, arXiv:0910.4301; BaBar, B. Aubert *et al.*, *Observation of the semileptonic decays $B \to D^*\tau^-\bar{\nu}(\tau)$*

*and evidence for* $B \to D\tau^-\bar{\nu}(\tau)$, Phys. Rev. Lett. **100** (2008) 021801, arXiv:0709.1698.

[241] A. K. Leibovich, Z. Ligeti, I. W. Stewart, and M. B. Wise, *Semileptonic B decays to excited charmed mesons*, Phys. Rev. D **57** (1998) 308, arXiv:hep-ph/9705467.

[242] CLEO collaboration, P. U. E. Onyisi *et al.*, *Improved measurement of absolute branching fraction of* $D_s^+ \to \tau^+\nu$, Phys. Rev. D **79** (2009) 052002, arXiv:0901.1147.

[243] Belle, S. Hirose *et al.*, *Measurement of the* $\tau$ *lepton polarization and* $R(D^*)$ *in the decay* $\bar{B} \to D^*\tau^-\bar{\nu}_\tau$ *with one-prong hadronic* $\tau$ *decays at Belle*, Phys. Rev. D **97** (2018) 012004, arXiv:1709.00129.

[244] M. Tanaka, *Charged Higgs effects on exclusive semitauonic B decays*, Z. Phys. C **67** (1995) 321, arXiv:hep-ph/9411405.

[245] M. Tanaka and R. Watanabe, $\tau$ *longitudinal polarization in* $B \to D\tau\nu$ *and its role in the search for charged Higgs boson*, Phys. Rev. D **82** (2010) 034027, arXiv:1005.4306.

[246] S. Fajfer, J. F. Kamenik, and I. Nisandzic, *On the* $B \to D^*\tau\bar{\nu}$ *Sensitivity to New Physics*, Phys. Rev. D **85** (2012) 094025, arXiv:1203.2654.

[247] A. Datta, M. Duraisamy, and D. Ghosh, *Diagnosing New Physics in* $b \to c\tau\nu$ *decays in the light of the recent BaBar result*, Phys. Rev. D **86** (2012) 034027, arXiv:1206.3760.

[248] P. Biancofiore, P. Colangelo, and F. De Fazio, *On the anomalous enhancement observed in* $B \to D^{(*)}\tau\bar{\nu}$ *decays*, Phys. Rev. D **87** (2013) 074010, arXiv:1302.1042.

[249] M. Tanaka and R. Watanabe, *New physics in the weak interaction of* $\bar{B} \to D^{(*)}\tau\bar{\nu}$, Phys. Rev. D **87** (2013) 034028, arXiv:1212.1878.

[250] Y. Sakaki, M. Tanaka, A. Tayduganov, and R. Watanabe, *Testing leptoquark models in* $\bar{B} \to D^{(*)}\tau\bar{\nu}$, Phys. Rev. D **88** (2013) 094012, arXiv:1309.0301.

[251] M. Duraisamy, P. Sharma, and A. Datta, *Azimuthal* $B \to D^*\tau^-\bar{\nu}_\tau$ *angular distribution with tensor operators*, Phys. Rev. D **90** (2014) 074013, arXiv:1405.3719.

[252] S. Bhattacharya, S. Nandi, and S. K. Patra, *Looking for possible new physics in* $B \to D^{(*)}\tau\nu_\tau$ *in light of recent data*, Phys. Rev. D **95** (2017) 075012, arXiv:1611.04605.

[253] D. Bardhan, P. Byakti, and D. Ghosh, *A closer look at the* $R_D$ *and* $R_{D^*}$ *anomalies*, JHEP **01** (2017) 125, arXiv:1610.03038.

[254] K. Hagiwara, A. D. Martin, and D. Zeppenfeld, *Tau Polarization Measurements at LEP and SLC*, Phys. Lett. B **235** (1990) 198.

[255] Belle, A. Abdesselam *et al.*, *Measurement of the* $D^{*-}$ *polarization in the decay* $B^0 \to D^{*-}\tau^+\nu_\tau$, in *10th International Workshop on the CKM Unitarity Triangle*, 2019, arXiv:1903.03102.

[256] Z.-R. Huang, Y. Li, C.-D. Lu, M. A. Paracha, and C. Wang, *Footprints of new physics in* $b \to c\tau\nu$ *transitions*, Phys. Rev. D **98** (2018) 095018, arXiv:1808.03565.

[257] S. Bhattacharya, S. Nandi, and S. Kumar Patra, $b \to c\tau\nu_\tau$ *decays: A catalogue to compare, constrain, and correlate new physics effects*, Eur. Phys. J. C **79** (2019) 268, arXiv:1805.08222.

[258] Belle, G. Caria *et al.*, *Measurement of* $\mathcal{R}(D)$ *and* $\mathcal{R}(D^*)$ *with a semileptonic tagging method*, Phys. Rev. Lett. **124** (2020) 161803, arXiv:1910.05864.

[259] LHCb, R. Aaij *et al.*, *Measurement of the shape of the* $B_s^0 \to D_s^{*-}\mu^+\nu_\mu$ *differential decay rate*, JHEP **12** (2020) 144, arXiv:2003.08453.

[260] LHCb, R. Aaij *et al.*, *First observation of the decay* $B_s^0 \to K^-\mu^+\nu_\mu$ *and measurement of* $|V_{ub}|/|V_{cb}|$, arXiv:2012.05143.

[261] LHCb, R. Aaij *et al.*, *Measurement of the ratio of branching fractions* $\mathcal{B}(\bar{B}^0 \to D^{*+}\tau^-\bar{\nu}_\tau)/\mathcal{B}(\bar{B}^0 \to D^{*+}\mu^-\bar{\nu}_\mu)$, Phys. Rev. Lett. **115** (2015) 111803, arXiv:1506.08614, [Erratum: Phys.Rev.Lett. 115, 159901 (2015)].

[262] LHCb, R. Aaij *et al.*, *First observation of* $\bar{B}_s^0 \to D_{s2}^{*+}X\mu^-\bar{\nu}$ *decays*, Phys. Lett. B **698** (2011) 14, arXiv:1102.0348.

[263] LHCb, R. Aaij *et al.*, *Measurement of b hadron fractions in 13 TeV pp collisions*, Phys. Rev. D **100** (2019) 031102, arXiv:1902.06794.

[264] LHCb collaboration, R. Aaij *et al.*, *Test of Lepton Flavor Universality by the measurement of the* $B^0 \to D^{*-}\tau^+\nu_\tau$ *branching fraction using three-prong* $\tau$ *decays*, Phys. Rev. D **97** (2018) 072013, arXiv:1711.02505.

[265] LHCb, R. Aaij *et al.*, *Measurement of the ratio of branching fractions* $\mathcal{B}(B_c^+ \to J/\psi\tau^+\nu_\tau)/\mathcal{B}(B_c^+ \to J/\psi\mu^+\nu_\mu)$, Phys. Rev. Lett. **120** (2018) 121801, arXiv:1711.05623.

[266] LHCb, R. Aaij *et al.*, *Measurement of the* $B_c^-$ *meson production fraction and asymmetry in 7 and 13 TeV pp collisions*, Phys. Rev. D **100** (2019) 112006, arXiv:1910.13404.

[267] F. U. Bernlochner, M. F. Sevilla, D. J. Robinson, and G. Wormser, *Semitauonic b-hadron decays: A lepton flavor universality laboratory*, Rev. Mod. Phys. **94** (2022) 1, arXiv:2101.08326.

[268] G. Isidori and O. Sumensari, *Optimized lepton universality tests in* $B \to V\ell\bar{\nu}$ *decays*, Eur. Phys. J. C **80** (2020) 1078, arXiv:2007.08481.

[269] HPQCD, C. Bouchard, G. P. Lepage, C. Monahan, H. Na, and J. Shigemitsu, *Rare decay* $B \to K\ell^+\ell^-$ *form factors from lattice QCD*, Phys. Rev. D **88** (2013) 054509, arXiv:1306.2384, [Erratum: Phys.Rev.D 88, 079901 (2013)].

[270] R. R. Horgan, Z. Liu, S. Meinel, and M. Wingate, *Lattice QCD calculation of form factors describing the rare decays* $B \to K^*\ell^+\ell^-$ *and* $B_s \to \phi\ell^+\ell^-$, Phys. Rev. D **89** (2014) 094501, arXiv:1310.3722.

[271] A. Bharucha, D. M. Straub, and R. Zwicky, $B \to V\ell^+\ell^-$ *in the Standard Model from light-cone sum rules*, JHEP **08** (2016) 098, arXiv:1503.05534.

[272] A. Khodjamirian, T. Mannel, and Y. M. Wang, $B \to K\ell^+\ell^-$ *decay at large hadronic recoil*, JHEP **02** (2013) 010, arXiv:1211.0234.

[273] C. Bobeth, M. Chrzaszcz, D. van Dyk, and J. Virto, *Long-distance effects in* $B \to K^*\ell\ell$ *from analyticity*, Eur. Phys. J. C **78** (2018) 451, arXiv:1707.07305.

[274] N. Gubernari, D. van Dyk, and J. Virto, *Non-local matrix elements in* $B_{(s)} \to \{K^{(*)}, \phi\}\ell^+\ell^-$, JHEP **02** (2021) 088, arXiv:2011.09813.

[275] M. Beneke, T. Feldmann, and D. Seidel, *Systematic approach to exclusive* $B \to Vl^+l^-$, $V\gamma$ *decays*, Nucl. Phys. B **612** (2001) 25, arXiv:hep-ph/0106067.

[276] M. Beneke, T. Feldmann, and D. Seidel, *Exclusive radiative and electroweak* $b \to d$ *and* $b \to s$ *penguin decays at NLO*, Eur. Phys. J. C **41** (2005) 173, arXiv:hep-ph/0412400.

[277] A. Ali, G. Kramer, and G.-h. Zhu, $B \to K^+l^+l^-$ *decay in soft-collinear effective theory*, Eur. Phys. J. C **47** (2006) 625, arXiv:hep-ph/0601034.

[278] C. Bobeth, G. Hiller, and G. Piranishvili, *Angular distributions of* $\bar{B} \to \bar{K}\ell^+\ell^-$ *decays*, JHEP **12** (2007) 040, arXiv:0709.4174.

[279] M. Bartsch, M. Beylich, G. Buchalla, and D.-N. Gao, *Precision Flavour Physics with* $B \to K\nu\bar{\nu}$ *and* $B \to Kl^+l^-$, JHEP **11** (2009) 011, arXiv:0909.1512.

[280] G. Isidori, S. Nabeebaccus, and R. Zwicky, *QED corrections in* $\overline{B} \to \overline{K}\ell^+\ell^-$ *at the double-differential level*, JHEP **12** (2020) 104, arXiv:2009.00929.

[281] G. Buchalla, A. J. Buras, and M. E. Lautenbacher, *Weak decays beyond leading logarithms*, Rev. Mod. Phys. **68** (1996) 1125, arXiv:hep-ph/9512380.

[282] A. J. Buras, J. Girrbach, D. Guadagnoli, and G. Isidori, *On the Standard Model prediction for* $\mathcal{B}(B_{s,d} \to \mu^+\mu^-)$, Eur. Phys. J. C **72** (2012) 2172, arXiv:1208.0934.

[283] C. Bobeth *et al.*, $B_{s,d} \to l^+l^-$ *in the Standard Model with Reduced Theoretical Uncertainty*, Phys. Rev. Lett. **112** (2014) 101801, arXiv:1311.0903.

[284] M. Beneke, C. Bobeth, and R. Szafron, *Enhanced electromagnetic correction to the rare B-meson decay* $B_{s,d} \to \mu^+\mu^-$, Phys. Rev. Lett. **120** (2018) 011801, arXiv:1708.09152.

[285] S. Descotes-Genon, J. Matias, and J. Virto, *Understanding the* $B \to K^*\mu^+\mu^-$ *Anomaly*, Phys. Rev. D **88** (2013) 074002, arXiv:1307.5683.

[286] W. Altmannshofer and D. M. Straub, *New Physics in* $B \to K^*\mu\mu$?, Eur. Phys. J. C **73** (2013) 2646, arXiv:1308.1501.

[287] T. Hurth and F. Mahmoudi, *On the LHCb anomaly in* $B \to K^*\ell^+\ell^-$, JHEP **04** (2014) 097, arXiv:1312.5267.

[288] G. Hiller and M. Schmaltz, $R_K$ *and future* $b \to s\ell\ell$ *physics beyond the standard model opportunities*, Phys. Rev. D **90** (2014) 054014, arXiv:1408.1627.

[289] R. Alonso, B. Grinstein, and J. Martin Camalich, $SU(2) \times U(1)$ *gauge invariance and the shape of new physics in rare B decays*, Phys. Rev. Lett. **113** (2014) 241802, arXiv:1407.7044.

[290] T. Hurth, F. Mahmoudi, and S. Neshatpour, *Global fits to* $b \to s\ell\ell$ *data and signs for lepton non-universality*, JHEP **12** (2014) 053, arXiv:1410.4545.

[291] M. Ciuchini *et al.*, *On Flavourful Easter eggs for New Physics hunger and Lepton Flavour Universality violation*, Eur. Phys. J. C **77** (2017) 688, arXiv:1704.05447.

[292] G. D'Amico *et al.*, *Flavour anomalies after the* $R_{K^*}$ *measurement*, JHEP **09** (2017) 010, arXiv:1704.05438.

[293] M. Algueró *et al.*, *Emerging patterns of New Physics with and without Lepton Flavour Universal contributions*, Eur. Phys. J. C **79** (2019) 714, arXiv:1903.09578, [Addendum: Eur.Phys.J.C 80, 511 (2020)].

[294] J. Aebischer *et al.*, *B-decay discrepancies after Moriond 2019*, Eur. Phys. J. C **80** (2020) 252, arXiv:1903.10434.

[295] M. Ciuchini *et al.*, *Lessons from the $B^{0,+} \to K^{*0,+}\mu^+\mu^-$ angular analyses*, Phys. Rev. D **103** (2021) 015030, arXiv:2011.01212.

[296] D. Lancierini, G. Isidori, P. Owen, and N. Serra, *On the significance of new physics in $b \to s\ell^+\ell^-$ decays*, Phys. Lett. B **822** (2021) 136644, arXiv:2104.05631.

[297] R. Barbieri, G. Isidori, J. Jones-Perez, P. Lodone, and D. M. Straub, *U(2) and minimal flavour violation in supersymmetry*, Eur. Phys. J. C **71** (2011) 1725, arXiv:1105.2296.

[298] R. Barbieri, D. Buttazzo, F. Sala, and D. M. Straub, *Flavour physics from an approximate $U(2)^3$ symmetry*, JHEP **07** (2012) 181, arXiv:1203.4218.

[299] G. Blankenburg, G. Isidori, and J. Jones-Perez, *Neutrino masses and LFV from minimal breaking of $U(3)^5$ and $U(2)^5$ flavor symmetries*, Eur. Phys. J. C **72** (2012) 2126, arXiv:1204.0688.

[300] C. Cornella, D. A. Faroughy, J. Fuentes-Martín, G. Isidori, and M. Neubert, *Reading the footprints of the B-meson flavor anomalies*, JHEP **08** (2021) 050, arXiv:2103.16558.

[301] L. Wolfenstein, *Violation of CP Invariance and the possibility of very weak interactions*, Phys. Rev. Lett. **13** (1964) 562.

[302] L. Di Luzio and M. Nardecchia, *What is the scale of new physics behind the B-flavour anomalies?*, Eur. Phys. J. C **77** (2017) 536, arXiv:1706.01868.

[303] D. Das, C. Hati, G. Kumar, and N. Mahajan, *Towards a unified explanation of $R_{D^{(*)}}$, $R_K$ and $(g-2)_\mu$ anomalies in a left-right model with leptoquarks*, Phys. Rev. D **94** (2016) 055034, arXiv:1605.06313.

[304] D. Das, C. Hati, G. Kumar, and N. Mahajan, *Scrutinizing R-parity violating interactions in light of $R_{K^{(*)}}$ data*, Phys. Rev. D **96** (2017) 095033, arXiv:1705.09188.

[305] S. Trifinopoulos, *Revisiting R-parity violating interactions as an explanation of the B-physics anomalies*, Eur. Phys. J. C **78** (2018) 803, arXiv:1807.01638.

[306] S. Trifinopoulos, *B-physics anomalies: The bridge between R-parity violating supersymmetry and flavored dark matter*, Phys. Rev. D **100** (2019) 115022, arXiv:1904.12940.

[307] W. Altmannshofer, P. S. B. Dev, A. Soni, and Y. Sui, *Addressing $R_{D^{(*)}}$, $R_{K^{(*)}}$, muon $g-2$ and ANITA anomalies in a minimal R-parity violating supersymmetric framework*, Phys. Rev. D **102** (2020) 015031, arXiv:2002.12910.

[308] R. Gauld, F. Goertz, and U. Haisch, *On minimal $Z'$ explanations of the $B \to K^*\mu^+\mu^-$ anomaly*, Phys. Rev. D **89** (2014) 015005, arXiv:1308.1959.

[309] A. J. Buras and J. Girrbach, *Left-handed $Z'$ and Z FCNC quark couplings facing new $b \to s\mu^+\mu^-$ data*, JHEP **12** (2013) 009, arXiv:1309.2466.

[310] W. Altmannshofer, S. Gori, M. Pospelov, and I. Yavin, *Quark flavor transitions in $L_\mu - L_\tau$ models*, Phys. Rev. D **89** (2014) 095033, arXiv:1403.1269.

[311] A. Crivellin, G. D'Ambrosio, and J. Heeck, *Addressing the LHC flavor anomalies with horizontal gauge symmetries*, Phys. Rev. D **91** (2015) 075006, arXiv:1503.03477.

[312] W. Altmannshofer and I. Yavin, *Predictions for lepton flavor universality violation in rare B decays in models with gauged $L_\mu - L_\tau$*, Phys. Rev. D **92** (2015) 075022, arXiv:1508.07009.

[313] B. Allanach, F. S. Queiroz, A. Strumia, and S. Sun, *$Z'$ models for the LHCb and $g - 2$ muon anomalies*, Phys. Rev. D **93** (2016) 055045, arXiv:1511.07447, [Erratum: Phys.Rev.D 95, 119902 (2017)].

[314] A. Crivellin, J. Fuentes-Martin, A. Greljo, and G. Isidori, *Lepton Flavor Non-Universality in B decays from Dynamical Yukawas*, Phys. Lett. B **766** (2017) 77, arXiv:1611.02703.

[315] B. C. Allanach and J. Davighi, *Third family hypercharge model for $R_{K^{(*)}}$ and aspects of the fermion mass problem*, JHEP **12** (2018) 075, arXiv:1809.01158.

[316] B. C. Allanach, J. M. Butterworth, and T. Corbett, *Collider constraints on $Z'$ models for neutral current B-anomalies*, JHEP **08** (2019) 106, arXiv:1904.10954.

[317] B. C. Allanach, *$U(1)_{B_3-L_2}$ explanation of the neutral current $B-$anomalies*, Eur. Phys. J. C **81** (2021) 56, arXiv:2009.02197, [Erratum: Eur. Phys. J. C 81, 321 (2021)].

[318] J. Davighi, *Anomalous $Z'$ bosons for anomalous B decays*, arXiv: 2105.06918.

[319] R. Alonso, B. Grinstein, and J. Martin Camalich, *Lepton universality violation and lepton flavor conservation in B-meson decays*, JHEP **10** (2015) 184, arXiv:1505.05164.

[320] V. Bernard, M. Oertel, E. Passemar, and J. Stern, *$K_{\mu 3}^L$ decay: A stringent test of right-handed quark currents*, Phys. Lett. B **638** (2006) 480, arXiv:hep-ph/0603202.

[321] V. Cirigliano, J. Jenkins, and M. Gonzalez-Alonso, *Semileptonic decays of light quarks beyond the Standard Model*, Nucl. Phys. B **830** (2010) 95, arXiv:0908.1754.

[322] R.-X. Shi, L.-S. Geng, B. Grinstein, S. Jäger, and J. Martin Camalich, *Revisiting the new-physics interpretation of the $b \to c\tau\nu$ data*, JHEP **12** (2019) 065, arXiv:1905.08498.

[323] J. Fuentes-Martín, G. Isidori, J. Pagès, and K. Yamamoto, *With or without $U(2)$? Probing non-standard flavor and helicity structures in semileptonic B decays*, Phys. Lett. B **800** (2020) 135080, arXiv:1909.02519.

[324] M. Freytsis, Z. Ligeti, and J. T. Ruderman, *Flavor models for $\bar{B} \to D^{(*)}\tau\bar{\nu}$*, Phys. Rev. D **92** (2015) 054018, arXiv:1506.08896.

[325] S. Fajfer and N. Košnik, *Vector leptoquark resolution of $R_K$ and $R_{D^{(*)}}$ puzzles*, Phys. Lett. B **755** (2016) 270, arXiv:1511.06024.

[326]  S. Bhattacharya, S. Nandi, and S. K. Patra, *Optimal-observable analysis of possible new physics in* $B \to D^{(*)}\tau\nu_\tau$, Phys. Rev. D **93** (2016) 034011, arXiv:1509.07259.

[327]  B. Bhattacharya, A. Datta, J.-P. Guévin, D. London, and R. Watanabe, *Simultaneous explanation of the* $R_K$ *and* $R_{D^{(*)}}$ *puzzles: A model analysis*, JHEP **01** (2017) 015, arXiv:1609.09078.

[328]  D. Becirevic, S. Fajfer, I. Nisandzic, and A. Tayduganov, *Angular distributions of* $\bar{B} \to D^{(*)}\ell\bar{\nu}_\ell$ *decays and search of New Physics*, Nucl. Phys. B **946** (2019) 114707, arXiv:1602.03030.

[329]  R. Alonso, A. Kobach, and J. Martin Camalich, *New physics in the kinematic distributions of* $\bar{B} \to D^{(*)}\tau^-(\to \ell^-\bar{\nu}_\ell\nu_\tau)\bar{\nu}_\tau$, Phys. Rev. D **94** (2016) 094021, arXiv:1602.07671.

[330]  X.-Q. Li, Y.-D. Yang, and X. Zhang, *Revisiting the one leptoquark solution to the* $R(D^{(*)})$ *anomalies and its phenomenological implications*, JHEP **08** (2016) 054, arXiv:1605.09308.

[331]  A. K. Alok, D. Kumar, S. Kumbhakar, and S. U. Sankar, $D^*$ *polarization as a probe to discriminate new physics in* $\bar{B} \to D^*\tau\bar{\nu}$, Phys. Rev. D **95** (2017) 115038, arXiv:1606.03164.

[332]  Z. Ligeti, M. Papucci, and D. J. Robinson, *New Physics in the visible final states of* $B \to D^{(*)}\tau\nu$, JHEP **01** (2017) 083, arXiv:1610.02045.

[333]  R. Alonso, B. Grinstein, and J. Martin Camalich, *Lifetime of* $B_c^-$ *constrains explanations for anomalies in* $B \to D^{(*)}\tau\nu$, Phys. Rev. Lett. **118** (2017) 081802, arXiv:1611.06676.

[334]  A. Celis, M. Jung, X.-Q. Li, and A. Pich, *Scalar contributions to* $b \to c(u)\tau\nu$ *transitions*, Phys. Lett. B **771** (2017) 168, arXiv:1612.07757.

[335]  M. Blanke *et al.*, *Impact of polarization observables and* $B_c \to \tau\nu$ *on new physics explanations of the* $b \to c\tau\nu$ *anomaly*, Phys. Rev. D **99** (2019) 075006, arXiv:1811.09603.

[336]  M. Jung and D. M. Straub, *Constraining new physics in* $b \to c\ell\nu$ *transitions*, JHEP **01** (2019) 009, arXiv:1801.01112.

[337]  C. Murgui, A. Peñuelas, M. Jung, and A. Pich, *Global fit to* $b \to c\tau\nu$ *transitions*, JHEP **09** (2019) 103, arXiv:1904.09311.

[338]  R. Mandal, C. Murgui, A. Peñuelas, and A. Pich, *The role of right-handed neutrinos in* $b \to c\tau\bar{\nu}$ *anomalies*, JHEP **08** (2020) 022, arXiv:2004.06726.

[339]  Belle, S. Hirose *et al.*, *Measurement of the* $\tau$ *lepton polarization and* $R(D^*)$ *in the decay* $\bar{B} \to D^*\tau^-\bar{\nu}_\tau$, Phys. Rev. Lett. **118** (2017) 211801, arXiv:1612.00529.

[340]  B. Bhattacharya, A. Datta, D. London, and S. Shivashankara, *Simultaneous Explanation of the* $R_K$ *and* $R(D^{(*)})$ *Puzzles*, Phys. Lett. B **742** (2015) 370, arXiv:1412.7164.

[341]  A. Greljo, G. Isidori, and D. Marzocca, *On the breaking of Lepton Flavor Universality in B decays*, JHEP **07** (2015) 142, arXiv:1506.01705.

[342]  L. Calibbi, A. Crivellin, and T. Ota, *Effective Field Theory Approach to* $b \to s\ell\ell^{(\prime)}$, $B \to K^{(*)}\nu\bar{\nu}$ *and* $B \to D^{(*)}\tau\nu$ *with Third Generation Couplings*, Phys. Rev. Lett. **115** (2015) 181801, arXiv:1506.02661.

[343] S. L. Glashow, D. Guadagnoli, and K. Lane, *Lepton Flavor Violation in B decays?*, Phys. Rev. Lett. **114** (2015) 091801, arXiv:1411.0565.

[344] R. Barbieri, G. Isidori, A. Pattori, and F. Senia, *Anomalies in B-decays and U(2) flavour symmetry*, Eur. Phys. J. C **76** (2016) 67, arXiv:1512.01560.

[345] M. Bordone, G. Isidori, and S. Trifinopoulos, *Semileptonic B-physics anomalies: A general EFT analysis within $U(2)^n$ flavor symmetry*, Phys. Rev. D **96** (2017) 015038, arXiv:1702.07238.

[346] B. Gripaios, M. Nardecchia, and S. A. Renner, *Composite leptoquarks and anomalies in B-meson decays*, JHEP **05** (2015) 006, arXiv:1412.1791.

[347] D. Bečirević, S. Fajfer, and N. Košnik, *Lepton flavor nonuniversality in $b \to s\ell^+\ell^-$ processes*, Phys. Rev. D **92** (2015) 014016, arXiv:1503.09024.

[348] M. Bauer and M. Neubert, *Minimal leptoquark explanation for the $R_{D^{(*)}}$, $R_K$, and $(g-2)_g$ anomalies*, Phys. Rev. Lett. **116** (2016) 141802, arXiv:1511.01900.

[349] D. Bečirević and O. Sumensari, *A leptoquark model to accommodate $R_K^{\text{exp}} < R_K^{\text{SM}}$ and $R_{K^*}^{\text{exp}} < R_{K^*}^{\text{SM}}$*, JHEP **08** (2017) 104, arXiv:1704.05835.

[350] D. Bečirević et al., *Scalar leptoquarks from grand unified theories to accommodate the B-physics anomalies*, Phys. Rev. D **98** (2018) 055003, arXiv:1806.05689.

[351] B. Diaz, M. Schmaltz, and Y.-M. Zhong, *The leptoquark Hunter's guide: Pair production*, JHEP **10** (2017) 097, arXiv:1706.05033.

[352] M. Schmaltz and Y.-M. Zhong, *The leptoquark Hunter's guide: Large coupling*, JHEP **01** (2019) 132, arXiv:1810.10017.

[353] I. Doršner and A. Greljo, *Leptoquark toolbox for precision collider studies*, JHEP **05** (2018) 126, arXiv:1801.07641.

[354] B. Gripaios, *Composite Leptoquarks at the LHC*, JHEP **02** (2010) 045, arXiv:0910.1789.

[355] F. Feruglio, P. Paradisi, and A. Pattori, *Revisiting Lepton Flavor Universality in B Decays*, Phys. Rev. Lett. **118** (2017) 011801, arXiv:1606.00524.

[356] F. Feruglio, P. Paradisi, and A. Pattori, *On the importance of electroweak corrections for B anomalies*, JHEP **09** (2017) 061, arXiv:1705.00929.

[357] D. A. Faroughy, A. Greljo, and J. F. Kamenik, *Confronting lepton flavor universality violation in B decays with high-$p_T$ tau lepton searches at LHC*, Phys. Lett. B **764** (2017) 126, arXiv:1609.07138.

[358] A. Greljo and D. Marzocca, *High-$p_T$ dilepton tails and flavor physics*, Eur. Phys. J. C **77** (2017) 548, arXiv:1704.09015.

[359] D. Buttazzo, A. Greljo, G. Isidori, and D. Marzocca, *B-physics anomalies: A guide to combined explanations*, JHEP **11** (2017) 044, arXiv:1706.07808.

[360] J. C. Pati and A. Salam, *Lepton Number as the Fourth Color*, Phys. Rev. D **10** (1974) 275, [Erratum: Phys.Rev.D 11, 703–703 (1975)].

[361] A. Crivellin, C. Greub, D. Müller, and F. Saturnino, *Importance of loop effects in explaining the accumulated evidence for new physics in B decays with a vector leptoquark*, Phys. Rev. Lett. **122** (2019) 011805, arXiv:1807.02068.

[362] C. Bobeth and U. Haisch, *New Physics in $\Gamma_{12}^s$: $(\bar{s}b)(\bar{\tau}\tau)$ Operators*, Acta Phys. Polon. B **44** (2013) 127, arXiv:1109.1826.

[363] M. Algueró, B. Capdevila, S. Descotes-Genon, P. Masjuan, and J. Matias, *Are we overlooking lepton flavour universal new physics in $b \to s\ell\ell$ ?*, Phys. Rev. D **99** (2019) 075017, arXiv:1809.08447.

[364] L. Di Luzio, J. Fuentes-Martin, A. Greljo, M. Nardecchia, and S. Renner, *Maximal Flavour Violation: a Cabibbo mechanism for leptoquarks*, JHEP **11** (2018) 081, arXiv:1808.00942.

[365] J. Fuentes-Martín, G. Isidori, M. König, and N. Selimović, *Vector leptoquarks beyond tree Level III: Vector-like fermions and flavor-changing transitions*, Phys. Rev. D **102** (2020) 115015, arXiv:2009.11296.

[366] V. Gherardi, D. Marzocca, and E. Venturini, *Matching scalar leptoquarks to the SMEFT at one loop*, JHEP **07** (2020) 225, arXiv:2003.12525, [Erratum: JHEP 01, 006 (2021)].

[367] V. Gherardi, D. Marzocca, and E. Venturini, *Low-energy phenomenology of scalar leptoquarks at one-loop accuracy*, JHEP **01** (2021) 138, arXiv:2008.09548.

[368] A. Angelescu, D. Bečirević, D. A. Faroughy, and O. Sumensari, *Closing the window on single leptoquark solutions to the B-physics anomalies*, JHEP **10** (2018) 183, arXiv:1808.08179.

[369] D. Marzocca, *Addressing the B-physics anomalies in a fundamental Composite Higgs Model*, JHEP **07** (2018) 121, arXiv:1803.10972.

[370] A. Greljo, D. J. Robinson, B. Shakya, and J. Zupan, *$R_D$ from $W'$ and right-handed neutrinos*, JHEP **09** (2018) 169, arXiv:1804.04642.

[371] G. Cvetič, F. Halzen, C. S. Kim, and S. Oh, *Anomalies in (semi)-leptonic B decays $B^{\pm} \to \tau^{\pm}\nu$, $B^{\pm} \to D\tau^{\pm}\nu$ and $B^{\pm} \to D^*\tau^{\pm}\nu$, and possible resolution with sterile neutrino*, Chin. Phys. C **41** (2017) 113102, arXiv:1702.04335.

[372] D. Bečirević, S. Fajfer, N. Košnik, and O. Sumensari, *Leptoquark model to explain the B-physics anomalies, $R_K$ and $R_D$*, Phys. Rev. D **94** (2016) 115021, arXiv:1608.08501.

[373] S. Fajfer, J. F. Kamenik, I. Nisandzic, and J. Zupan, *Implications of lepton flavor universality violations in B decays*, Phys. Rev. Lett. **109** (2012) 161801, arXiv:1206.1872.

[374] L. Di Luzio, M. Kirk, and A. Lenz, *Updated $B_s$-mixing constraints on new physics models for $b \to s\ell^+\ell^-$ anomalies*, Phys. Rev. D **97** (2018) 095035, arXiv:1712.06572.

[375] HFLAV, Y. Amhis *et al.*, *Averages of b-hadron, c-hadron, and τ-lepton properties as of summer 2016*, Eur. Phys. J. C **77** (2017) 895, arXiv: 1612.07233.

[376] LHCb, R. Aaij *et al.*, *Search for the decays $B_s^0 \to \tau^+\tau^-$ and $B^0 \to \tau^+\tau^-$*, Phys. Rev. Lett. **118** (2017) 251802, arXiv:1703.02508.

[377] BaBar, J. P. Lees *et al.*, *Search for $B^+ \to K^+\tau^+\tau^-$ at the BaBar experiment*, Phys. Rev. Lett. **118** (2017) 031802, arXiv:1605.09637.

[378] LHCb, R. Aaij *et al.*, *Search for the lepton-flavour-violating decays $B_s^0 \to \tau^{\pm}\mu^{\mp}$ and $B^0 \to \tau^{\pm}\mu^{\mp}$*, Phys. Rev. Lett. **123** (2019) 211801, arXiv:1905.06614.

[379] BaBar, J. P. Lees *et al.*, *A search for the decay modes* $B^{\pm} \to h^{\pm}\tau^{\mp}\ell^{\pm}$, Phys. Rev. D **86** (2012) 012004, arXiv:1204.2852.

[380] Belle, Y. Miyazaki *et al.*, *Search for Lepton-Flavor-Violating* $\tau$ *decays into a lepton and a vector meson*, Phys. Lett. B **699** (2011) 251, arXiv:1101.0755.

[381] J. Blumlein, E. Boos, and A. Kryukov, *Leptoquark pair production in hadronic interactions*, Z. Phys. C **76** (1997) 137, arXiv:hep-ph/9610408.

[382] M. J. Baker, J. Fuentes-Martín, G. Isidori, and M. König, *High- $p_T$ signatures in vector–leptoquark models*, Eur. Phys. J. C **79** (2019) 334, arXiv:1901.10480.

[383] CMS, A. M. Sirunyan *et al.*, *Search for singly and pair-produced leptoquarks coupling to third-generation fermions in proton-proton collisions at* $\sqrt{s} =$ *13 TeV*, arXiv:2012.04178.

[384] A. Angelescu, D. Bečirević, D. A. Faroughy, F. Jaffredo, and O. Sumensari, *On the single leptoquark solutions to the B-physics anomalies*, Phys. Rev. D **104** (2021) 055017, arXiv:2103.12504.

[385] ATLAS, G. Aad *et al.*, *Search for heavy Higgs bosons decaying into two tau leptons with the ATLAS detector using pp collisions at* $\sqrt{s} = 13$ *TeV*, Phys. Rev. Lett. **125** (2020) 051801, arXiv:2002.12223.

[386] A. Angelescu, D. A. Faroughy, and O. Sumensari, *Lepton Flavor Violation and Dilepton Tails at the LHC*, Eur. Phys. J. C **80** (2020) 641, arXiv:2002.05684.

[387] A. Greljo, J. Martin Camalich, and J. D. Ruiz-Álvarez, *Mono-$\tau$ signatures at the LHC constrain explanations of B-decay anomalies*, Phys. Rev. Lett. **122** (2019) 131803, arXiv:1811.07920.

[388] U. Haisch and G. Polesello, *Resonant third-generation leptoquark signatures at the Large Hadron Collider*, JHEP **05** (2021) 057, arXiv:2012.11474.

[389] A. Greljo and N. Selimovic, *Lepton-quark fusion at hadron colliders, precisely*, JHEP **03** (2021) 279, arXiv:2012.02092.

[390] J. Kumar, D. London, and R. Watanabe, *Combined Explanations of the* $b \to s\mu^+\mu^-$ *and* $b \to c\tau^-\bar{\nu}$ *Anomalies: A General Model Analysis*, Phys. Rev. D **99** (2019) 015007, arXiv:1806.07403.

[391] M. Bordone, C. Cornella, G. Isidori, and M. König, *The LFU ratio* $R_\pi$ *in the Standard Model and beyond*, Eur. Phys. J. C **81** (2021) 9, arXiv:2101.11626.

[392] R. Barbieri, C. W. Murphy, and F. Senia, *B-decay anomalies in a composite leptoquark model*, Eur. Phys. J. C **77** (2017) 8, arXiv:1611.04930.

[393] R. Barbieri and A. Tesi, *B-decay anomalies in Pati-Salam SU(4)*, Eur. Phys. J. C **78** (2018) 193, arXiv:1712.06844.

[394] G. F. Giudice, G. Isidori, A. Salvio, and A. Strumia, *Softened gravity and the extension of the Standard Model up to infinite energy*, JHEP **02** (2015) 137, arXiv:1412.2769.

[395] N. Assad, B. Fornal, and B. Grinstein, *Baryon number and lepton universality violation in leptoquark and diquark models*, Phys. Lett. B **777** (2018) 324, arXiv:1708.06350.

[396] L. Calibbi, A. Crivellin, and T. Li, *Model of vector leptoquarks in view of the B-physics anomalies*, Phys. Rev. D **98** (2018) 115002, arXiv:1709.00692.

[397]  L. Di Luzio, A. Greljo, and M. Nardecchia, *Gauge leptoquark as the origin of B-physics anomalies*, Phys. Rev. D **96** (2017) 115011, arXiv:1708.08450.

[398]  M. Bordone, C. Cornella, J. Fuentes-Martin, and G. Isidori, *A three-site gauge model for flavor hierarchies and flavor anomalies*, Phys. Lett. B **779** (2018) 317, arXiv:1712.01368.

[399]  B. Fornal, S. A. Gadam, and B. Grinstein, *Left-right SU(4) vector leptoquark model for flavor anomalies*, Phys. Rev. D **99** (2019) 055025, arXiv:1812.01603.

[400]  M. Blanke and A. Crivellin, *B meson anomalies in a Pati-Salam model within the Randall-Sundrum background*, Phys. Rev. Lett. **121** (2018) 011801, arXiv:1801.07256.

[401]  A. Greljo and B. A. Stefanek, *Third family quark–lepton unification at the TeV scale*, Phys. Lett. B **782** (2018) 131, arXiv:1802.04274.

[402]  C. Cornella, J. Fuentes-Martin, and G. Isidori, *Revisiting the vector leptoquark explanation of the B-physics anomalies*, JHEP **07** (2019) 168, arXiv:1903.11517.

[403]  J. Fuentes-Martín and P. Stangl, *Third-family quark-lepton unification with a fundamental composite Higgs*, Phys. Lett. B **811** (2020) 135953, arXiv:2004.11376.

[404]  H. Georgi and Y. Nakai, *Diphoton resonance from a new strong force*, Phys. Rev. D **94** (2016) 075005, arXiv:1606.05865.

[405]  J. Fuentes-Martín, G. Isidori, J. Pagès, and B. A. Stefanek, *Flavor non-universal Pati-Salam unification and neutrino masses*, arXiv:2012.10492.

[406]  J. Fuentes-Martín, G. Isidori, M. König, and N. Selimović, *Vector Leptoquarks Beyond Tree Level*, Phys. Rev. D **101** (2020) 035024, arXiv:1910.13474.

[407]  J. Fuentes-Martín, G. Isidori, M. König, and N. Selimović, *Vector leptoquarks beyond tree level. II. $\mathcal{O}(\alpha_s)$ corrections and radial modes*, Phys. Rev. D **102** (2020) 035021, arXiv:2006.16250.

[408]  J. Heeck and D. Teresi, *Pati-Salam explanations of the B-meson anomalies*, JHEP **12** (2018) 103, arXiv:1808.07492.

[409]  P. F. Perez, C. Murgui, and A. D. Plascencia, *Leptoquarks and matter unification: Flavor anomalies and the muon g − 2*, Phys. Rev. D **104** (2021) 035041, arXiv:2104.11229.

[410]  A. Greljo, P. Stangl, and A. E. Thomsen, *A model of muon anomalies*, Phys. Lett. B **820** (2021) 136554, arXiv:2103.13991.

[411]  Muon g − 2, B. Abi et al., *Measurement of the positive muon anomalous magnetic moment to 0.46 ppm*, Phys. Rev. Lett. **126** (2021) 141801, arXiv:2104.03281.

[412]  D. Marzocca and S. Trifinopoulos, *Minimal Explanation of Flavor Anomalies: B-Meson Decays, Muon Magnetic Moment, and the Cabibbo Angle*, Phys. Rev. Lett. **127** (2021) 061803, arXiv:2104.05730.

[413]  S. Baum, M. Carena, N. R. Shah, and C. E. M. Wagner, *The Tiny (g − 2) Muon Wobble from Small-μ Supersymmetry*, JHEP **01** (2022) 025, arXiv:2104.03302.

[414] R. Barbieri, *A View of Flavour Physics in 2021*, Acta Phys. Polon. B **52** (2021) 789, arXiv:2103.15635.

[415] L. Allwicher, G. Isidori, and A. E. Thomsen, *Stability of the Higgs sector in a flavor-inspired multi-scale model*, JHEP **01** (2021) 191, arXiv:2011.01946.

[416] G. Panico and A. Pomarol, *Flavor hierarchies from dynamical scales*, JHEP **07** (2016) 097, arXiv:1603.06609.

[417] G. R. Dvali and M. A. Shifman, *Families as neighbors in extra dimension*, Phys. Lett. B **475** (2000) 295, arXiv:hep-ph/0001072.

[418] A. Crivellin *et al.*, *Lepton-flavour violating B decays in generic Z' models*, Phys. Rev. D **92** (2015) 054013, arXiv:1504.07928.

[419] I. de Medeiros Varzielas and G. Hiller, *Clues for flavor from rare lepton and quark decays*, JHEP **06** (2015) 072, arXiv:1503.01084.

[420] M. Duraisamy, S. Sahoo, and R. Mohanta, *Rare semileptonic B → K(π)ℓ_i^- ℓ_j^+ decay in a vector leptoquark model*, Phys. Rev. D **95** (2017) 035022, arXiv:1610.00902.

[421] D. Bečirević, O. Sumensari, and R. Zukanovich Funchal, *Lepton flavor violation in exclusive b → s decays*, Eur. Phys. J. C **76** (2016) 134, arXiv:1602.00881.

[422] D. Bečirević, N. Košnik, O. Sumensari, and R. Zukanovich Funchal, *Palatable leptoquark scenarios for Lepton Flavor Violation in exclusive b → sℓ_1ℓ_2 modes*, JHEP **11** (2016) 035, arXiv:1608.07583.

[423] M. Bordone, C. Cornella, J. Fuentes-Martín, and G. Isidori, *Low-energy signatures of the PS³ model: From B-physics anomalies to LFV*, JHEP **10** (2018) 148, arXiv:1805.09328.

[424] J. Alda, J. Guasch, and S. Penaranda, *Some results on Lepton Flavour Universality Violation*, Eur. Phys. J. C **79** (2019) 588, arXiv:1805.03636.

[425] A. D. Smirnov, *Vector leptoquark mass limits and branching ratios of K_L^0, B^0, B_s → ℓ_i^+ ℓ_j^- decays with account of fermion mixing in leptoquark currents*, Mod. Phys. Lett. A **33** (2018) 1850019, arXiv:1801.02895.

[426] CLEO, R. Ammar *et al.*, *Search for B^0 decays to two charged leptons*, Phys. Rev. D **49** (1994) 5701.

[427] CLEO, A. Bornheim *et al.*, *Search for the lepton-flavor-violating leptonic B decays B^0 → μ^±τ^∓ and B^0 → e^±τ^∓*, Phys. Rev. Lett. **93** (2004) 241802, arXiv:hep-ex/0408011.

[428] BaBar, B. Aubert *et al.*, *Searches for the decays B^0 → ℓ^±τ^∓ and B^+ → ℓ^+ν (l=e, μ) using hadronic tag reconstruction*, Phys. Rev. D **77** (2008) 091104, arXiv:0801.0697.

[429] S. Stone and L. Zhang, *Method of Studying Λ_b^0 decays with one missing particle*, Adv. High Energy Phys. **2014** (2014) 931257, arXiv:1402.4205.

[430] LHCb, R. Aaij *et al.*, *Search for lepton-flavor violating decays B^+ → K^+μ^±e^∓*, Phys. Rev. Lett. **123** (2019) 241802, arXiv:1909.01010.

[431] BaBar, B. Aubert *et al.*, *Measurements of branching fractions, rate asymmetries, and angular distributions in the rare decays B → Kℓ^+ℓ^- and B → K^*ℓ^+ℓ^-*, Phys. Rev. D **73** (2006) 092001, arXiv:hep-ex/0604007.

[432] GERDA, M. Agostini *et al.*, *Final results of GERDA on the search for neutrinoless double-β Decay*, Phys. Rev. Lett. **125** (2020) 252502, arXiv:2009.06079.

[433]  EXO-200, G. Anton *et al.*, *Search for neutrinoless double-β decay with the complete EXO-200 dataset*, Phys. Rev. Lett. **123** (2019) 161802, arXiv:1906.02723.

[434]  LHCb collaboration, R. Aaij *et al.*, *Searches for Majorana neutrinos in $B^-$ decays*, Phys. Rev. D **85** (2012) 112004, arXiv:1201.5600.

[435]  BELLE, O. Seon *et al.*, *Search for lepton-number-violating $B^+ \to D^- l^+ l'^+$ Decays*, Phys. Rev. D **84** (2011) 071106, arXiv:1107.0642.

[436]  A. Atre, T. Han, S. Pascoli, and B. Zhang, *The search for heavy Majorana neutrinos*, JHEP **05** (2009) 030, arXiv:0901.3589.

[437]  J.-M. Zhang and G.-L. Wang, *Lepton-number violating decays of heavy mesons*, Eur. Phys. J. C **71** (2011) 1715, arXiv:1003.5570.

[438]  A. J. Weir *et al.*, *Upper limits on $D^\pm$ and $B^\pm$ decays to two leptons plus $\pi^\pm$ or $K^\pm$*, Phys. Rev. D **41** (1990) 1384.

[439]  CLEO, K. W. Edwards *et al.*, *Search for lepton flavor violating decays of B mesons*, Phys. Rev. D **65** (2002) 111102, arXiv:hep-ex/0204017.

[440]  LHCb, R. Aaij *et al.*, *Search for the lepton number violating decays $B^+ \to \pi^- \mu^+ \mu^+$ and $B^+ \to K^- \mu^+ \mu^+$*, Phys. Rev. Lett. **108** (2012) 101601, arXiv:1110.0730.

[441]  Belle, D. Liventsev *et al.*, *Search for heavy neutrinos at Belle*, Phys. Rev. D **87** (2013) 071102, arXiv:1301.1105, [Erratum: Phys.Rev.D 95, 099903 (2017)].

[442]  LHCb collaboration, R. Aaij *et al.*, *Search for Majorana neutrinos in $B^- \to \pi^+ \mu^- \mu^-$ decays*, Phys. Rev. Lett. **112** (2014) 131802, arXiv:1401.5361.

[443]  B. Shuve and M. E. Peskin, *Revision of the LHCb limit on Majorana neutrinos*, Phys. Rev. D **94** (2016) 113007, arXiv:1607.04258.

[444]  N. Quintero, G. Lopez Castro, and D. Delepine, *Lepton number violation in top quark and neutral B meson decays*, Phys. Rev. D **84** (2011) 096011, arXiv:1108.6009, [Erratum: Phys.Rev.D 86, 079905 (2012)].

[445]  A. Ali, A. V. Borisov, and N. B. Zamorin, *Majorana neutrinos and same sign dilepton production at LHC and in rare meson decays*, Eur. Phys. J. C **21** (2001) 123, arXiv:hep-ph/0104123; G. Cvetic, C. Dib, S. K. Kang, and C. S. Kim, *Probing Majorana neutrinos in rare K and D, $D_s$, B, $B_c$ meson decays*, Phys. Rev. D **82** (2010) 053010, arXiv:1005.4282.

[446]  N. Quintero, *Constraints on lepton number violating short-range interactions from $|\Delta L| = 2$ processes*, Phys. Lett. B **764** (2017) 60, arXiv:1606.03477.

[447]  S. Cunliffe, *The Belle II experiment: First results, status, and prospects*, https://docs.belle2.org/record/2202?ln=en, 2021.

[448]  P. Langacker, *The standard model and beyond*, CRC Press, 2010.

[449]  C. Quigg, *Gauge theories of the strong, weak, and electromagnetic Interactions: Second Edition*, Princeton University Press, USA, 2013.

[450]  W. N. Cottingham and D. A. Greenwood, *An introduction to the standard model of particle physics*, Cambridge University Press, 2007. The SM Lagrangian can be found here: https://www.einstein-schrodinger.com/Standard_Model.pdf.

[451] Y. Grossman, *Introduction to flavor physics*, in *2009 European School of High-Energy Physics*, 111–144, 2010, arXiv:1006.3534.

[452] Y. Nir, *Flavour physics and CP violation*, in *7th CERN–Latin-American School of High-Energy Physics*, 123–156, 2015, arXiv:1605.00433.

[453] E. Majorana, *Teoria simmetrica dell'elettrone e del positrone*, Nuovo Cim. **14** (1937) 171.

[454] A. Strumia and F. Vissani, *Neutrino masses and mixings and...*, arXiv:hep-ph/0606054.

[455] J. M. Gerard, *Fermion mass spectrum in SU(2)-L x U(1)*, Z. Phys. C **18** (1983) 145.

[456] R. S. Chivukula and H. Georgi, *Composite technicolor Standard Model*, Phys. Lett. B **188** (1987) 99.

[457] C. Bobeth, M. Gorbahn, and E. Stamou, *Electroweak corrections to $B_{s,d} \to \ell^+\ell^-$*, Phys. Rev. D **89** (2014) 034023, arXiv:1311.1348.

[458] M. Gorbahn, U. Haisch, and M. Misiak, *Three-loop mixing of dipole operators*, Phys. Rev. Lett. **95** (2005) 102004, arXiv:hep-ph/0504194.

[459] C. Bobeth, P. Gambino, M. Gorbahn, and U. Haisch, *Complete NNLO QCD analysis of $\overline{B} \to X_s\ell^+\ell^-$ and higher order electroweak effects*, JHEP **04** (2004) 071, arXiv:hep-ph/0312090.

[460] K. G. Chetyrkin, M. Misiak, and M. Munz, *Weak radiative B meson decay beyond leading logarithms*, Phys. Lett. B **400** (1997) 206, arXiv:hep-ph/9612313, [Erratum: Phys.Lett.B 425, 414 (1998)].

[461] M. Ciuchini, E. Franco, G. Martinelli, and L. Reina, *$\epsilon'/\epsilon$ at the Next-to-leading order in QCD and QED*, Phys. Lett. B **301** (1993) 263, arXiv:hep-ph/9212203.

[462] A. J. Buras, M. Jamin, M. E. Lautenbacher, and P. H. Weisz, *Two loop anomalous dimension matrix for $\Delta S = 1$ weak nonleptonic decays I: $\mathcal{O}(\alpha_s^2)$*, Nucl. Phys. B **400** (1993) 37, arXiv:hep-ph/9211304.

[463] T. Blake, G. Lanfranchi, and D. M. Straub, *Rare B decays as tests of the Standard Model*, Prog. Part. Nucl. Phys. **92** (2017) 50, arXiv:1606.00916.

[464] M. González-Alonso, J. Martin Camalich, and K. Mimouni, *Renormalization-group evolution of new physics contributions to (semi) leptonic meson decays*, Phys. Lett. B **772** (2017) 777, arXiv:1706.00410.

[465] M. Bordone, M. Jung, and D. van Dyk, *Theory determination of $\bar{B} \to D^{(*)}\ell^-\bar{\nu}$ form factors at $\mathcal{O}(1/m_c^2)$*, Eur. Phys. J. C **80** (2020) 74, arXiv:1908.09398.

www.ingramcontent.com/pod-product-compliance
Lightning Source LLC
Chambersburg PA
CBHW050558190326
41458CB00007B/2093